Quantum Biology

The Strange Physics Powering Life

Team Trex

Published by NobleTrex Press

For permissions and other inquiries, write to:

P.O. Box 3132, Framingham, MA 01701, USA

Contents

Introduction

Consider the European robin for a moment. It is a small, round creature, barely weighing twenty grams-roughly the same as a handful of coins or a slightly overfilled letter. To the casual observer, it is a garden ornament, a splash of red breast flitting between hedgerows. But every autumn, this fragile ball of feathers undertakes a journey that defies imagination. It launches itself from the chilling forests of Scandinavia and navigates thousands of miles south to the warmth of the Mediterranean or North Africa. It flies primarily by night, over featureless oceans and unfamiliar terrain, maintaining a course so precise that it can return to the exact same branch the following spring.

For decades, biologists assumed the robin navigated like a Boy Scout, relying on visual landmarks or the position of the stars. Later, they realized the bird possessed an internal compass sensitive to the Earth's magnetic field. But here lies the problem: the Earth's magnetic field is incredibly weak. It is significantly weaker than the magnetic field emanating from your refrigerator door. A biological sensor capable of detecting such a faint signal amidst the noisy chemical environment of a living cell faces a signal-to-noise challenge that would baffle our best engineers.

How does the robin do it? The answer, according to a growing body of rigorous evidence, is that the bird is not merely doing biology. It is doing quantum physics.

Inside the robin's eye, specifically within a protein called cryptochrome, a photon of blue light strikes a molecule, knocking an electron loose. This creates a pair of "radicals"-molecules with unpaired electrons. These two electrons are separated in space, yet they remain intimately connected by the spooky phenomenon known as quantum entanglement. For a fleeting microsecond, they dance in a delicate balance, spinning in a way that makes them hypersensitive to the angle of the Earth's magnetic field lines. The bird, quite literally, "sees" the magnetic field as a pattern of light and shadow overlaying its vision.

If this sounds like science fiction, you are not alone in your skepticism. For the better part of a century, the idea that the warm, wet, and messy environment of a living organism could sustain delicate quantum states was considered scientific heresy. Quantum mechanics-the physics of the very small-was supposed to be the domain of hermetically sealed vacuums, absolute zero temperatures, and laser-stabilized laboratories. In the chaotic bustle of a cell, where molecules are constantly colliding and vibrating with thermal energy, quantum effects like coherence and entanglement should be destroyed almost instantly. This destruction is called "decoherence," and it is the primary reason we don't see people walking through walls or being in two places at once.

And yet, life seems to have missed the memo.

We are standing at the threshold of a new scientific frontier: Quantum Biology. This is not a fringe discipline of mysticism or vague energy fields. It is a hard science, forged at the intersection of molecular biology, physical chemistry, and quantum mechanics. It asks a profound and unsettling question: Does life merely exist on top of the laws of physics, or has evolution actually learned to exploit the strangest aspects of the quantum world to gain a survival advantage?

The evidence suggests the latter. And the robin is just the beginning.

Walk outside and look at a green leaf. It is performing the most important chemical reaction on Earth: photosynthesis. It captures a photon of sunlight and transfers that energy to a reaction center where it can be stored as food. If you were to design a machine to do this using classical physics, you would expect significant energy loss. The energy packet (the exciton) would bounce around randomly like a drunkard in a hallway, losing heat with every collision, hoping to stumble upon the exit. But plants don't do that. In many photosynthetic systems, the energy transfer occurs with an efficiency bordering on 95 to 99 percent.

How? Experiments suggest that the energy doesn't hop from molecule to molecule. Instead, it enters a state of superposition. Like a ripple in a pond, the energy wave travels all possible paths simultaneously. It explores every route to the reaction center at once, inevitably finding the most efficient path in a fraction of a second. The humblest moss on a sidewalk is performing a quantum computation that our most advanced supercomputers struggle to simulate.

This book is an exploration of that hidden layer of reality. We are going to peel back the skin of the world we think we know to look at the machinery humming underneath. We are not just talking about birds and leaves; we are talking about you.

Deep inside your own body, enzymes are currently breaking down your breakfast, facilitating chemical reactions that, according to classical thermodynamics, should take years to occur on their own. Life accelerates these reactions by factors of trillions. While the "lock and key" model of enzymes is taught in every high school biology class, it fails to account for the sheer speed of these processes. It turns out that protons-the nuclei of hydrogen atoms-are likely "tunneling" through energy barriers. They are vanishing from one position and instantly materializing in another, bypassing the energy hill they are supposed to climb. Your metabolism, the very fire of your existence, may depend on subatomic particles walking through walls.

As we delve into these phenomena, we must navigate carefully. The word "quantum" has unfortunately become a marketing buzzword, often slapped onto everything from consciousness theories to healing crystals to make them sound scientific. This book is an antidote to that hype. We will be rigorous in distinguishing between what is known, what is hypothesized, and what is wishful thinking. We will look at the hard data-the spectroscopic signatures, the magnetic field disruptions, and the isotopic labeling-that separates the signal from the noise.

One of the most fascinating themes we will uncover is how life handles the "warm, wet problem." Physicists spend millions of dollars cooling their quantum computers to near absolute zero to protect them from environmental noise. Life, however, seems to have adopted a judo-like strategy. Rather than fighting the thermal noise, biological systems appear to use it. We will explore the concept of "Goldilocks" vibrations-how a specific amount of molecular shaking actually helps maintain quantum coherence rather than destroying it. It turns out that life thrives on the edge of chaos, utilizing the very disorder that should kill it to drive its engines.

Our investigation begins by looking backward. We will revisit the prescient thoughts of Erwin Schrödinger, the Austrian physicist famous for his hypothetical cat. In 1944, long before the discovery of the double helix, Schrödinger wrote a small book titled *What is Life?* in which he argued that the stability of genetic material across generations-an "aperiodic crystal"-could only be explained by quantum mechanical stability. He was right, of course, but he couldn't have foreseen that the dynamics of life, not just its structure, would be equally indebted to the quantum realm.

From the foundations of genetics, we will move to the "impossible toolkit"-a primer on the strange rules of the quantum road. If you have ever felt intimidated by the concept of wave-particle duality or entanglement, fear not. We will strip away the dense mathematics and focus on the conceptual beauty of these ideas. You don't need to solve the Schrödinger equation to understand how a particle can explore two paths at once, any more than you need to understand internal combustion to drive a car.

Once we have our toolkit, we will travel into the heart of the green world. We will meet the "Green Bacteria Paradox" and see how ancient organisms solved the energy transport problem billions of years before humans invented the wheel. We will examine the controversy of "quantum beats"-the rhythmic pulsing observed in labs that suggests nature is humming a quantum tune.

Then, we will turn our gaze to the animal kingdom. Beyond the robin, we will look at the olfactory sense. There is a controversial but compelling theory that our sense of smell relies not just on the shape of molecules, but on their vibrations-essentially, that our noses are listening to the song of smell via electron tunneling. We will look at the implications for medicine, wondering if the efficacy of certain drugs or anesthetics relies on quantum properties we have ignored for centuries.

Finally, we will look to the future. If nature has been building quantum computers and quantum sensors for three billion years, what can we learn from her? The emerging field of quantum biology is already inspiring a new generation of solar cells that mimic the architecture of leaves. It is driving the development of navigation systems that don't rely on satellites. It is forcing us to reconsider the definition of life itself.

Is life simply complex chemistry? Or is it something more fundamental? If quantum mechanics is the rulebook of the universe, and biology is the most complex player in the game, it stands to reason that the player has mastered the rules.

The story of quantum biology is a story of convergence. For a hundred years, physics and biology were like two neighbors who lived in the same building but never spoke. Biologists were busy cataloging the immense complexity of living forms, while physicists were obsessed with the fundamental simplicity of matter. Now, the wall between the apartments is coming down. We are realizing that to understand the macro, we must understand the micro. To understand the flight of the bird, we must understand the spin of the electron.

This is not a journey into the mystical. It is a journey into the mechanism. It is a realization that the "strange" physics of the quantum world is not confined to the laboratory. It is happening in your garden, in your eyes, and in the very enzymes knitting your body together as you read this sentence.

Welcome to the edge of reality. Welcome to the strange, vibrating, impossible physics that powers life.

Chapter 1

Life on the Edge of Chaos

While physicists usually trap atoms in vacuums near absolute zero to observe quantum behavior, life thrives in a hot, chaotic soup of molecular collisions. We begin by examining Erwin Schrödinger's daring 1944 prediction that life must rely on a "quantum code" to resist thermodynamic decay. This opens the door to the central paradox of quantum biology: how fragile quantum states survive the thermal storm of the living cell long enough to perform useful work.

1.1 Schrödinger's Prophecy

In the early 1940s, biology was full of confident descriptions and missing mechanisms. Cells could be stained and sketched. Chromosomes could be watched, in a limited way, as they lined up and separated. Mutations could be inferred from changed traits. Yet heredity itself remained strangely ghostly. What, physically, was a gene? How could something so small store a specification for an organism and then copy that specification again and again, with an error rate low enough to keep a species recognizable over centuries?

Erwin Schrödinger asked that question with a physicist's impatience. By then he was famous for the wave equation that helped make quantum mechanics calculable, and for a habit of turning hard problems into sharp ones. In 1944, in a short book based on public lectures, *What Is Life?*, he tried to state the puzzle of heredity in a form that could not be dodged: living things maintain a high degree of internal order, and they preserve patterns across generations, despite being made of matter that is constantly jostled by heat. If an organism is warm, wet, and alive, every piece of it is being shaken by thermal motion; the molecules in its tissues are being knocked about by surrounding molecules, as if by an invisible hailstorm. How, in that setting, does a stable message persist?

A part of the answer was already implicit in chemistry: strong covalent bonds can make robust molecules. Another part would come later, with the structure of DNA. Schrödinger's originality was not in guessing a double helix he could not have seen. It was in seeing that heredity posed a problem of *information stored in matter*, and that any plausible solution had to be stable for reasons deeper than mere averaging.

He framed the issue by contrasting two kinds of order. One is the kind that statistical physics had already made familiar: order that appears *on average* when vast numbers of particles behave collectively. A gas has pressure and temperature not because any single molecule is special, but because so many molecules are there that their individual chaos washes out into a reliable bulk law. Even a crystal, at first glance, seems to fit this picture: it is orderly because the same pattern repeats in a regular lattice, and the repetition makes the whole structure resilient.

But genes are not supposed to be repetitive in that way. A salt crystal is ordered because it is the same thing over and over again. A hereditary molecule must be ordered while still being able to say *different* things. It must carry a message with a potentially vast vocabulary: eye colour, enzyme specificity, body plan. Repetition would be the enemy. A gene cannot be a simple periodic crystal, like a three-dimensional wallpaper. It must be something else: an object with a stable structure, yet one that can vary in a controlled, combinatorial way.

Schrödinger's term for this hypothetical object was the "aperiodic crystal." The phrase is almost self-contradictory, and that is part of its power. A crystal suggests rigid stability, a structure that holds its shape against thermal agitation because its atoms occupy definite positions in a low-energy arrangement. Aperiodic suggests a lack of repeating unit, a sequence that can encode arbitrary patterns. Schrödinger wanted both at once: *order without repetition.* He imagined a molecule that is solid-like at the scale that matters for copying, but whose detailed arrangement along its length can vary to represent a code.

Today, with DNA so familiar that it appears on coffee mugs and courtroom exhibits, it is easy to miss how daring this was. In 1944, the leading candidates for hereditary material were often proteins, because proteins looked complicated enough to do the job. DNA, by contrast, seemed too monotonous: a polymer made from only four similar building blocks. Schrödinger's aperiodic crystal did not insist on DNA, but it did insist on the underlying logic: heredity must be carried by a *molecule* with a stable, discrete structure whose variations are not blurred away by heat.

That insistence was, at bottom, a quantum claim. Schrödinger did not mean that organisms must keep themselves cold, or that genes rely on mystical quantum entanglement across the body. He meant something more modest and more profound: the stability of matter at the molecular scale is itself a quantum achievement.

Classical physics, if taken literally, leaves atoms and molecules in an awkward position. In the classical picture, electrons orbit nuclei like planets around a sun, and accelerating charges should radiate energy. The orbit should decay; the atom should collapse. The reason matter does not collapse is that electrons do not follow classical orbits. They occupy quantized states, standing-wave-like patterns described by quantum mechanics. These patterns have discrete energies. The lowest one, the ground state, is stable; an electron cannot radiate away energy continuously because it cannot occupy energies in between. The same quantization that makes atoms stable also makes chemical bonds robust. Covalent bonds arise from quantum wavefunctions that spread over multiple atoms, lowering energy and creating preferred geometries. In other words, a molecule can be *solid* because quantum mechanics permits stable, discrete arrangements of electrons and nuclei.

Schrödinger's aperiodic crystal was a way of saying: if heredity is real, it must be written in a medium whose reliability is not statistical but structural. A message that lasts requires something like the digital stability of a printed page. The ink molecules do not have to be periodically arranged; the letters on the page certainly are not periodic. Yet the page is stable because the molecules in the ink and paper sit in quantum-defined bonds and configurations that do not melt into randomness at room temperature. Schrödinger was describing the gene, in effect, as a molecular text: a "code-script" whose permanence depends on the discreteness of molecular structure.

This was also how he approached the thermodynamic unease that haunted biology. The second law of thermodynamics is often paraphrased as the tendency of disorder to increase. More carefully, it says that in an isolated system, entropy tends to rise. Living organisms are not isolated; they are open systems, exchanging energy and matter with their environment. Yet the spectacle of life is still striking: organisms build and maintain exquisitely ordered structures, from cell membranes to bones to neural circuits. Schrödinger popularized a phrase that has sometimes been misunderstood: organisms, he said, "feed on negative entropy." He was not proposing a new substance. He was pointing out that to maintain internal order, an organism must export disorder to its surroundings, paying for local organization with increased entropy elsewhere. In modern language, organisms consume free energy and dissipate heat; they maintain low internal entropy by pushing entropy out.

The link to heredity is subtle but crucial. Metabolism can keep a system away from equilibrium for a while, but heredity demands something more: a way to *preserve a specification* over time. You can keep a candle flame going by supplying wax and oxygen, but the flame's shape is not a stable code; it is a dynamic pattern that depends continuously on conditions. A gene, by contrast, is a pattern that can be copied, repaired, and transmitted, and the copying must be accurate enough that the pattern survives. Schrödinger was pushing biology away from purely analog, fluid pictures and toward a view in which the essential instructions are stored in a discrete, molecular form.

A good historical anecdote makes this more concrete. When Schrödinger delivered those lectures, the world was at war. Many scientists had been displaced; ideas travelled in strange ways. In Cambridge and elsewhere, young researchers in biology read *What Is Life?* and felt, as later memoirs would put it, a door open. The book did not contain experimental protocols. It did not provide a catalogue of genes. It offered something rarer: a statement that the deepest problems of life were not immune to physics. For a generation that had watched quantum mechanics and relativity transform the understanding of nature, this was an invitation. Francis Crick would later acknowledge that Schrödinger's book helped draw him toward biology. Whatever one thinks of the book's specific guesses, its role as intellectual propaganda for a new kind of molecular biology is hard to deny.

Yet Schrödinger's prophecy was not simply "genes are molecules." Many people already suspected that. The deeper prophecy was that heredity depends on *quantum discreteness*. The classical world, in which properties vary continuously and small perturbations accumulate, is a poor substrate for faithful copying. A code that persists must be built from elements that can be distinguished reliably: one base rather than another, one arrangement rather than its neighbour. Quantum mechanics is the reason that matter comes with such discrete options. Atoms have specific valences. Bonds have preferred lengths and angles. Rotations around bonds are constrained by energy barriers. The landscape of possible configurations is not smooth but structured into minima and barriers. That structure is what makes a molecule a plausible memory device.

To see why this matters, imagine a purely classical gene. It would be some kind of tiny mechanical pattern whose details could slide continuously. Copying it would be like copying a soft wax seal while it is being warmed: each copy would smear slightly, and the smearing would accumulate. Natural selection can tolerate some errors, and indeed mutations are the raw material of evolution. But the baseline has to be stable enough that an organism's basic instructions are not overwritten by thermal noise each generation. Schrödinger's point was that the gene must be protected from the tyranny of continuous drift. The stability must be "law-like," enforced by the physics of the molecule itself.

This is also why the aperiodic crystal idea anticipates a modern theme: the connection between information and physical implementation. In contemporary terms, DNA is a polymer whose monomers are chosen from an alphabet of four bases. The sequence is a discrete string. Copying involves templated base pairing, a process that relies on the geometry and energetics of hydrogen bonds and stacking interactions, all of which are quantum-chemical in origin. Errors occur, but they are not the default; the chemistry is biased toward correct pairing, and cells invest further energy in proofreading and repair. Schrödinger did not know these details. He did not need to. His claim was that without some such discrete, stable encoding, heredity would be miraculous.

The "crystal" part of the phrase matters as much as the "aperiodic." Schrödinger wanted solidity: a structure that can exist in a small number of stable states, resistant to the buffeting of thermal motion. Crystals in the everyday sense have long-range periodicity, but the more important feature is that they are *locked* into a configuration by energy barriers. If you nudge a crystal at room temperature, it does not rearrange its internal pattern; it vibrates around fixed positions. Schrödinger imagined genes as similarly locked, not by repetition, but by a robust internal architecture. In modern molecular terms, this corresponds to the fact that DNA's sugar-phosphate backbone provides structural regularity, while the sequence of bases provides variability. The regularity supplies a reliable scaffold; the variability carries the message.

There is a temptation, when talking about quantum biology, to treat quantum mechanics as an exotic add-on to chemistry, like a rare spice. Schrödinger's prophecy points in a different direction. Quantum mechanics is not an optional flourish; it is the reason chemistry exists as a stable set of rules. If life is built from molecules, it is already built from quantum matter. The genuine question is not whether biology is "quantum" in some vague sense. The question is whether living systems exploit quantum phenomena in ways that go beyond the baseline quantum chemistry that makes molecules stable at all.

Heredity sits at an interesting boundary. On one hand, the stability of genetic material is almost entirely a story of quantum chemistry:

bonds, energy landscapes, and discrete configurations. On the other hand, the act of copying DNA is a dynamic process in a warm environment, full of fluctuations, and it involves choices that could, in principle, be sensitive to quantum effects at the level of individual protons or electrons. Schrödinger did not spell out such mechanisms, but he set up the physical problem that later researchers would tackle with greater precision: how can a system remain orderly and information-rich while being constantly kicked by thermal motion?

The aperiodic crystal also carried an implicit rebuke to a certain kind of vitalism. Vitalism was the old idea that living matter requires a special life force beyond physics and chemistry. By the 1940s it was already losing ground, but it lingered in the sense that life might be too complex or too purposeful to be reduced to physical law. Schrödinger's move was to reverse the burden of proof. If hereditary stability exists, he argued, then physics must be able to account for it. If classical physics seems inadequate, that is not an invitation to mysticism; it is a clue about the relevant scale. At the molecular scale, quantum mechanics governs what structures are possible and how stable they are. The gene, in this view, is not a violation of physics but an advertisement for the kind of physics that matters.

A final, often overlooked feature of Schrödinger's prophecy is its humility about what counts as an explanation. To say "DNA stores information" is not yet to explain how information is stored. It is to name a location. Schrödinger demanded a physical mechanism: a structure stable enough to hold a code, and flexible enough to carry a rich message. The aperiodic crystal is a concept that sits between physics and biology: it is simultaneously a material object and an informational one. That dual nature, so commonplace now in talk of genomes and codes, was then a radical way to think.

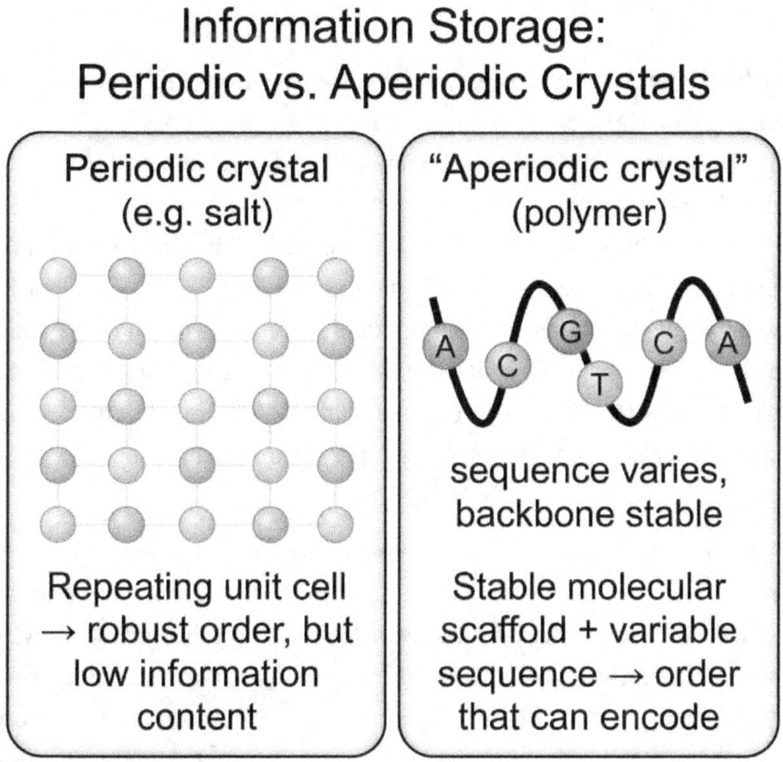

What makes this prophecy feel strangely modern is that it locates life's mystery in a narrow gap: between the wash of thermal agitation and the crispness of molecular discreteness. Schrödinger's gene is not a ghost in the machine. It is an object that must survive being shaken, because its stability is built into the allowed states of matter. In that sense, every faithful copy of hereditary information is a quiet demonstration of quantum rules: not the spectacular quantum of paradox and cat-states, but the quantum of solid, dependable architecture, where a particular arrangement can endure long enough to be read, replicated, and handed on.

1.2 The Thermal Storm

Schrödinger's "aperiodic crystal" was a wager on stability: a molecular script rigid enough to persist, yet flexible enough to encode. That wager works because chemical bonds are not easily blurred into half-states; they are protected by quantized energy levels and by energy barriers that are tall compared with ordinary thermal jitters. The deeper wager of quantum biology is riskier. It asks whether life sometimes does more than merely *endure* quantum rules at the molecular scale. It asks whether living systems can sometimes *use* distinctly quantum behaviours—coherence, tunnelling, spin dynamics—in places where heat seems determined to erase them.

The obstacle is not subtle. A living cell is not an isolated jewel of matter. It is a crowded bath at something like room temperature, soaked in water, salted with ions, filled with proteins that flex, twist, and collide. Even the language biologists use gives the game away: "cytoplasm" sounds like a gentle medium; in reality it behaves more like a jammed molecular city, where every object is nudged by the objects around it. If quantum mechanics is associated in the public mind with pristine laboratories, vacuum chambers, and delicate lasers, the cell looks like the wrong venue.

Physicists have a phrase for this discomfort: the "warm, wet, and noisy" problem. Warm means thermal energy is abundant. Wet means strong coupling to a solvent, especially water, which is polar and eager to interact. Noisy means ceaseless fluctuations: molecules bumping, dipoles reorienting, charges screening and rescreening, proteins breathing in and out. All of these features are excellent for chemistry—they keep molecules mobile enough to react—but they are poison for many quantum effects that depend on maintaining a precise phase relationship.

The relevant word is *decoherence*. It is often conflated with "collapse of the wavefunction", but the more useful idea is less metaphysical and more physical. Decoherence is what happens when a quantum system becomes entangled with its environment so thoroughly that, for practical purposes, the system can no longer display interference. The system has not necessarily done anything dramatic; it has merely leaked information into its surroundings. And in a cell, there are endless surroundings.

A simple way to picture coherence is to think of a wave with a well-defined rhythm. Interference is what you get when rhythms combine: peaks reinforcing peaks, troughs canceling troughs, patterns emerging that cannot be produced by a classical mixture of alternatives. Quantum coherence is the capacity of a system to behave like that wave: to keep track of relative phase between different possibilities. Once that relative phase is scrambled—by collisions, by fluctuating electromagnetic fields, by vibration, by random shifts in energy levels—the interference patterns wash out. The system then behaves like a classical lottery: it is in one state or another with certain probabilities, but it cannot *interfere* with itself to produce distinctly quantum signatures.

It is tempting to treat decoherence as something that happens *after* a quantum event, as if the cell permits a brief moment of pure quantum behaviour and then rudely interrupts it. In practice, decoherence is typically happening *during* the process of interest. The quantum system is not a sealed box; it is an *open quantum system*, exchanging energy and information with its environment. That matters because most of the quantum phenomena that sound useful for biology—efficient energy transfer, exquisitely selective reactions, delicate magnetic sensing—occur on timescales that compete with the timescales of environmental agitation. If the environment "checks" the system too frequently, it repeatedly nudges the system back into classicality.

Even the phrase "checks" risks suggesting an observer with intentions. But the environment need not be conscious, or even complicated. A single water molecule bumping into a biomolecule can act, in effect, like a measurement. It does not record the result in a notebook; it carries away a trace in its own state—a tiny change in momentum, orientation, or vibrational excitation. Multiply that by trillions of collisions per second, and the environment becomes an army of uninvited auditors.

The cell's violence can be expressed in a number that is both mundane and decisive: kT. Here k is Boltzmann's constant and T is temperature. At room temperature, kT is about 25 millielectronvolts, an energy scale that tells you how big thermal fluctuations typically are. That does not sound like much until you remember how small the energy gaps can be in biomolecular processes. If two quantum states differ in energy by an amount comparable to kT, then thermal motion can jostle the system between them, and random fluctuations can blur the phase relationship between them. One can think of kT as the budget the environment has to disrupt your carefully arranged quantum story.

Time is the other ruthless parameter. In a liquid, the interval between molecular collisions is often on the order of femtoseconds to picoseconds (10^{-15} to 10^{-12} seconds). Proteins and other biomolecules also possess dense forests of vibrational modes. These vibrations are not mere background; they are active participants, shifting electronic energies and coupling to charges and dipoles. If a quantum degree of freedom—say, an electron shared coherently across two sites—needs to retain phase coherence for, say, a nanosecond (10^{-9} seconds) to matter biologically, it must survive a huge number of environmental kicks.

It is worth pausing on why this is not simply a technical difficulty but a conceptual one. Quantum mechanics is universal: everything is quantum, in principle. Decoherence is the bridge between that universality and the apparent classical world. In a certain sense, the cell is not an exception to quantum theory; it is a poster child for why classical behaviour is so common. The very features that make biology lively—temperature, water, motion—also make quantum interference hard to detect.

A historical anecdote helps fix the mood. In the late twentieth century, as quantum information and quantum computation became fashionable, some thinkers began to wonder whether the brain might be a quantum computer in disguise. One of the most influential cold showers came from Max Tegmark (1999), who estimated decoherence times for various proposed neural-scale quantum states. The details depend on modeling choices, but the rhetorical impact was clear: in a warm biological medium, macroscopic quantum superpositions are expected to decohere fantastically fast, often far faster than the millisecond timescales on which neurons operate. The point was not that quantum mechanics is irrelevant to brains—ion channels and neurotransmitters are chemistry, and chemistry is quantum—but that the particular kind of long-lived, large-scale coherence imagined in some proposals is extremely difficult to sustain in such conditions.

That episode is useful because it separates two questions that are often blurred. One question is whether life *contains* quantum phenomena. It does; everything made of atoms does. The harder question is whether life maintains and exploits *fragile* quantum resources that are usually destroyed by environmental coupling. Tegmark's analysis was aimed at this second question, and it illustrates why physicists feel entitled to skepticism when someone waves a hand and says "quantum" as if the word itself conjures an explanation.

Yet skepticism should not harden into dogma. Decoherence is not a single number; it is a process, and it depends on what is decohering, how it couples to its environment, and what the environment is like.

A biomolecule is not a dust grain in air, and a cell is not a featureless heat bath. The environment inside a protein can be structured. Motions can be correlated. Solvent can be partly excluded. The coupling can be strong but selective. Much of modern quantum biology lives in that nuance: the question is not whether the thermal storm exists, but whether there are sheltered alleys within it.

To see what is at stake, it helps to distinguish two related ways the environment interferes with quantum behaviour. One is *energy relaxation*: the system loses energy to its surroundings, settling into lower-energy states. The other is *dephasing*: the system loses phase coherence without necessarily losing energy. Dephasing is often the immediate enemy of interference. A system can retain its energy and yet lose the ability to interfere if random fluctuations shift its phase at unpredictable rates.

A familiar laboratory comparison makes the point without romanticizing it. Superconductors can support quantum coherence over macroscopic distances, but they generally require cold temperatures. Cooling reduces thermal agitation and suppresses many decohering processes. In contrast, a biomolecule at physiological temperature cannot be cooled without killing the organism. If coherence appears in biology, it must do so without the crutch of refrigeration. That does not mean coherence must be long-lived in an absolute sense; it might be enough for it to persist for tens or hundreds of femtoseconds if the biological process of interest happens on that timescale. But it does mean the system must accomplish its task quickly, or else find clever ways to reduce sensitivity to noise.

Noise itself is not always purely destructive. A counterintuitive idea that will matter later is that intermediate noise can sometimes help transport processes by preventing certain kinds of quantum localization or trapping. This is not a loophole that makes the thermal storm friendly; it is more like the discovery that a well-timed gust can help a sailboat that would otherwise be stuck in a dead zone. Too little noise can leave the system stuck in patterns of destructive interference; too much noise ruins coherence entirely. The cell cannot choose "no noise," so if quantum effects are relevant, they likely live in a compromise regime where noise is shaped, not eliminated.

The analogy to measurement can be sharpened with a picture. Imagine a wavefunction as a smooth ripple spreading across a small region of space. In a perfectly isolated system, the ripple can remain smooth and coherent, with well-defined phase relationships across its extent. Now place that ripple in water, not as a metaphor but as a real physical setting: a watery environment full of dipoles and collisions. Each collision imprints a tiny mark. Some collisions shift the energy levels, others couple different states, others entangle the system with vibrational modes. The ripple becomes ragged. After enough impacts, what remains is not a smooth interference-capable wave but a patchwork that behaves like a classical mixture.

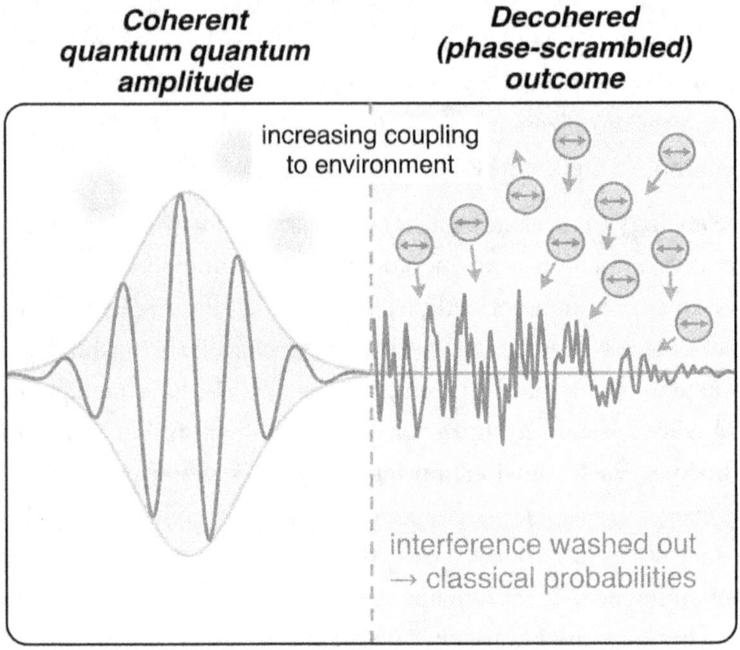

Thermal motion (water, ions, protein vibrations) continually entangles with the system, scrambling phase and suppressing quantum interference: decoherence.

The figure is only a cartoon, but it highlights a real asymmetry. Coherence is not a substance you possess; it is a relationship you must maintain. The environment does not need to do anything clever to destroy it. It only needs to interact.

This is why claims about quantum effects in biology must always come with a timescale and a location. *Where* is the quantum degree of freedom, exactly? An electron delocalized over a few pigment molecules? A proton that can tunnel between two positions in an enzyme active site? A pair of electron spins whose relative state encodes magnetic information? *For how long* must the relevant coherence persist? A femtosecond is unimaginably short in human terms, but it is long enough for some molecular events. A microsecond is short in neural terms but long enough for many chemical processes. Without this bookkeeping, "quantum" becomes an applause line rather than an explanation.

A further complication is that the environment in biology is not only noisy; it is also *functional*. Proteins are not rigid. They fluctuate, and those fluctuations are often essential for their role. A protein that never moved would be useless as an enzyme; it must breathe to bind substrates, to rearrange charges, to lower reaction barriers. Water is not merely a nuisance; it is the medium that permits diffusion, that stabilizes ions, that participates in reactions. The very features that help life operate classically can also be the features that destroy the more delicate quantum behaviours.

Modern discussions of decoherence sometimes sound as if it is a simple exponential decay: coherence disappears with a single characteristic time constant, like the discharge of a capacitor. That can be a good approximation in some regimes, but it can also mislead. Real environments can have memory. Their fluctuations can be correlated in time. In physics terms, the environment may be non-Markovian: its present behaviour can retain traces of its past. That matters because short-time behaviour can deviate from the simplest exponential story, and because correlated noise can sometimes be less destructive than uncorrelated noise. A structured environment can, in effect, push on a system in a coordinated way rather than delivering purely random kicks.

None of this removes the basic message of the thermal storm. It only sharpens it. If quantum effects matter in biology, they will do so in spite of the environment, or by making the environment part of the mechanism. Either way, the cell cannot be treated as a silent stage on which a quantum drama is performed. The cell is an active participant, a restless audience that coughs, shifts in its seat, and occasionally storms the stage.

A concrete example makes the violence visceral. Consider an electron that might, in isolation, occupy a superposition of being on site A and site B. In the cell, those sites are embedded in a protein scaffold surrounded by water and ions. The electron's charge slightly polarizes its surroundings: water dipoles rotate a little, nearby charges shift. If the electron is on A, the environment is polarized one way; if it is on B, another way. When the electron is in a superposition, the combined system becomes entangled: "electron on A" correlates with "environment adjusted to A," and "electron on B" correlates with "environment adjusted to B." If those two environmental states become distinguishable—even slightly—then the electron's coherence between A and B is reduced when you ignore the environment (which, in practice, you must). The environment has, in effect, learned which site the electron occupies, and that information is enough to suppress interference.

That is the basic reason the cell seems such an unlikely home for quantum coherence. It is not that quantum mechanics fails in warm systems; it is that warmth supplies many channels by which information leaks away.

And yet life persists on that knife-edge between rigidity and agitation. The aperiodic crystal survives because its essential degrees of freedom are locked by strong bonds and high barriers, while other degrees of freedom—vibrations, rotations, conformational shifts—remain lively. The possibility that matters for quantum biology is that some of those lively degrees of freedom might be arranged so that, for a brief moment and in a small place, the storm does not entirely wash out quantum structure. In a world of constant collisions, the remarkable thing would not be that coherence dies, but that any coherent whisper can be heard at all.

1.3 Beyond the Mysticism

Quantum mechanics has a peculiar cultural afterlife. It began as an austere theory of spectra and scattering experiments, a toolbox for calculating what atoms and electrons will do. Somewhere along the way it acquired a second identity: a kind of intellectual fog machine, capable of making any claim look profound if the lighting is right. Add the words "quantum" and "energy" to almost anything—healing, leadership, consciousness, love—and there will be an audience willing to nod.

Biology, because it already feels miraculous, is especially vulnerable to this temptation. A fertilized egg becomes a baby; a brain produces a mind; a bird migrates across oceans and returns to the same grove of trees. When a phenomenon is emotionally charged and mechanistically complicated, people reach for explanations that sound powerful. Quantum mechanics is powerful, and it is also weird. The combination is irresistible.

The trouble is that *weird* is not a scientific criterion. In the laboratory, quantum mechanics is not a license to believe in the impossible; it is a discipline of extraordinarily strict accounting. You must specify what system you are talking about, what quantum property matters, how long it lasts, what measurements reveal it, and why a classical explanation is insufficient. In other words, quantum biology has to be more precise than ordinary biology, not less, because it lives in a regime where sloppy language can masquerade as insight.

A practical boundary begins with a simple question: *Which quantum resource is doing the work?* The word "quantum" is not a single ingredient. It can mean at least four different things, and confusing them is one of the commonest routes to mysticism.

One meaning is *quantization*: energies and states come in discrete packets. This is the reason chemistry is reliable, the reason bonds have characteristic strengths and geometries, and the reason Schrödinger's genetic "script" could exist at all. It is profound, but it is also ubiquitous. If a claim about biology amounts to "it uses chemical bonds," then the quantum content is real but unremarkable.

A second meaning is *tunnelling*: a particle crosses an energy barrier it would not surmount in classical physics, because quantum mechanics allows a probability amplitude to leak through. Tunnelling is not magic; it has a calculable dependence on barrier height and width. It is also common in chemistry. Whether it is *functionally important* in a particular enzyme or receptor is a separate, testable question.

A third meaning is *coherence*: a system occupies a superposition of alternatives with well-defined relative phase, enabling interference. Coherence can be fleeting in warm matter. If coherence is invoked, timescales matter. A coherent effect lasting 300 femtoseconds might be biologically relevant in a process that happens in picoseconds; it is irrelevant to cognition unfolding over hundreds of milliseconds.

A fourth meaning is *entanglement*: correlations that cannot be reproduced by classical probabilities. Entanglement often accompanies coherence in multi-part systems, but it is not the same thing. It is also not automatically useful. Entanglement can appear briefly and do nothing functionally. Invoking it as a synonym for "spooky connection" is usually a warning sign.

Once the quantum resource is named, a second question follows: *What is the subsystem that carries it?* In a living organism, you cannot treat "the cell" or "the brain" as a single coherent quantum object unless you can also explain how it avoids the thermal storm. The more macroscopic the proposed quantum state, the more brutal decoherence becomes. This is the point made sharply by Tegmark's 1999 analysis of decoherence times for neural-scale superpositions:

the difficulty is not that quantum mechanics fails, but that the environment monitors large-scale states extremely efficiently. For claims about consciousness, this is often the decisive mismatch: the proposed quantum behaviour must persist at scales and times that are, on straightforward physical grounds, implausible.

Then comes the third question: *What is the environment model?* The environment is not an inconvenience to be waved away. In open quantum systems, the environment determines what decoheres and how fast. If a claim assumes that a biomolecule is isolated in something like a vacuum, it is not a claim about biology. Conversely, if a claim does take the environment seriously—a protein pocket that screens water, vibrational modes that are correlated, a solvent with finite memory—then the physics may become subtle, and the possibility of nontrivial quantum behaviour becomes an empirical matter rather than an article of faith.

A fourth question pins the discussion down: *What falsifiable consequence follows?* "Quantum effects make it more efficient" is not enough unless you can say what would change if the effect were absent. Would reaction rates shift with isotope substitution? Would a spectroscopic signature disappear? Would behaviour change under a controlled perturbation? Science advances when there is something specific to be wrong about.

These questions may sound like bureaucratic hurdles, but they are, in practice, the difference between a serious quantum-biological proposal and a motivational poster.

One way to see the border between rigorous and mystical is to look at how genuine quantum biology earned its credibility. It did not do so by claiming that life violates physics. It did so by making contact with measurable quantities: spectra, reaction rates, magnetic-field dependencies, isotope effects. When researchers argue about whether a given oscillation in a spectroscopy experiment reflects electronic coherence or vibronic mixing, they are arguing within a shared discipline of evidence. The debate can be heated precisely because it is anchored in data.

Compare that with the typical pseudoscientific use of quantum language. The claim is often elastic: when pressed for details, it becomes metaphor. When asked for a mechanism, it becomes an appeal to mystery. When confronted with a negative result, it becomes an argument that the effect is "too subtle" for ordinary measurement. That is not quantum physics; it is a strategy for avoiding being wrong.

A concrete example is helpful because it shows how close one can get to the border without crossing into fantasy. Consider smell. The everyday experience of olfaction is intensely qualitative and emotionally loaded. A faint whiff can summon childhood in an instant. Because of that, smell has attracted grand theories, including some that sound delightfully quantum.

The mainstream view, supported by a great deal of molecular biology, is that olfactory receptors detect odorants largely by their shape and chemical features: what fits, what binds, what triggers a conformational change and downstream signalling. In the 1990s and 2000s, an alternative called "vibrational" olfaction gained attention, sometimes associated with Luca Turin. The idea, in one influential form, is that receptors might detect not only shape but also the vibrational frequencies of a molecule, perhaps via inelastic electron tunnelling. Inelastic tunnelling is a real quantum process: an electron tunnels while exchanging energy with a vibrational mode. In principle, that could provide a spectral sensitivity of sorts.

It is an ingenious proposal, and it has the virtue of being testable. A molecule and its isotopomer—the same chemical structure but with some atoms replaced by heavier isotopes, such as hydrogen replaced by deuterium—have very similar shapes and chemistry but different vibrational frequencies. If receptors are vibration-sensitive, then isotopic substitution might sometimes change perceived smell, even when the overall geometry is nearly unchanged.

Some behavioural experiments have reported discrimination between isotopomers in insects, including Drosophila, which is intriguing. It suggests that under some circumstances, isotopic substitution can matter to olfaction. But this is where the discipline of evidence matters. Isotopic substitution changes more than vibrations; it can subtly change binding kinetics, zero-point energies, and interactions with the environment. A behavioural result does not automatically identify the mechanism. It sets a puzzle, not a conclusion.

When similar ideas have been tested in humans under controlled psychophysics, the results have been far less supportive of strong vibration-theory predictions. And at the receptor level, in vitro tests of specific receptors have sometimes failed to show isotopomer discrimination where vibrational theories might have expected it. That does not prove that vibrational effects never matter; it shows how hard it is to claim a quantum mechanism without pinning down the molecular details. Olfaction remains scientifically fascinating precisely because it resists simple stories, not because it invites mystical ones.

This example also reveals a common category error. People hear "quantum tunnelling" and imagine something like telepathy: information passing through barriers without contact. In reality, tunnelling is local and constrained. Its probability falls off rapidly with barrier width. It does not let a system ignore chemistry; it *is* chemistry, calculated in a different regime. A quantum effect in a receptor, if it exists, will not be a free-floating psychic antenna. It will be a particular electron moving between particular sites with particular couplings, producing a measurable change in reaction yield or signalling rate.

The consciousness debate provides an even starker illustration of boundary-setting. It is easy to see why quantum consciousness proposals flourish. Consciousness feels unified and private; quantum mechanics includes non-classical correlations; therefore, the temptation is to connect them. But the gap between suggestive analogy and workable mechanism is enormous. A brain operates in a warm, wet environment with relentless decoherence. Neural signalling is slow on the scale of molecular motion. Whatever quantum coherence exists at the microscopic level is rapidly converted into classical outcomes at the macroscopic level. To claim that consciousness depends on long-lived, large-scale quantum coherence is to propose not merely a new neuroscience but a new method of maintaining coherence in the harshest possible medium.

That is not logically impossible. Nature occasionally surprises. But the evidential burden is correspondingly heavy. One would need to identify a specific physical substrate, show that it sustains coherence on relevant timescales, demonstrate that this coherence is functionally necessary for neural computation, and explain how it is protected against decoherence. In the absence of such a chain of evidence, quantum consciousness remains less a scientific hypothesis than a reflection of how quantum mechanics makes people feel.

A more honest and more interesting stance is that biology does not need quantum mysticism because it already contains real quantum puzzles that are hard enough. The cell's challenge is not to evade physics but to operate at the edge where physics becomes delicate. Thermal energy is large enough to rattle molecular systems, yet small enough that energy gaps and barriers still matter. Interactions with the environment are strong enough to cause rapid decoherence, yet structured enough that some quantum effects may persist fleetingly.

This is a regime in which careful models of open quantum dynamics, grounded in spectroscopy and kinetics, are genuinely required.

That brings us to what the field means when it is being careful. Quantum biology, in its rigorous form, does not claim that organisms are quantum computers. It claims that certain biological functions may depend on quantum phenomena that are both (i) nontrivial relative to ordinary equilibrium chemistry, and (ii) demonstrably present under biologically relevant conditions. The emphasis is on *local mechanisms*. An electron transfer step in an enzyme. An exciton moving across a photosynthetic complex. A radical pair of electron spins influenced by Earth's magnetic field. Each of these is small enough that quantum theory is the natural language, and each occurs quickly enough that even short-lived coherence can, in principle, matter.

There is also a cultural boundary to draw, not just an intellectual one. The mystified use of quantum language often thrives on the notion that science is a set of secret permissions: words that, once spoken, override ordinary causality. The real power of quantum mechanics is the opposite. It is a constraint. It tells you what cannot be built, and what can only be built under strict conditions. Decoherence is not a philosophical complaint; it is a quantitative limit. Energy scales are not vibes; they are numbers. When the numbers are wrong, the story is wrong.

A historical detail from the early days of molecular biology captures the spirit. Before DNA's structure was known, many researchers felt that heredity was too complex to be a simple molecule. Schrödinger's aperiodic crystal pushed against that intuition by insisting on the physicality of the code. That insistence was not mystical; it was reductive in the best sense. It demanded that heredity be *made of something* stable enough to be copied. The subsequent success of molecular genetics did not eliminate wonder; it relocated wonder into the details: base pairing, replication fidelity, repair enzymes, the astonishing choreography of transcription and translation. The wonder survived because it was pinned to mechanisms.

The boundary between science and mysticism is not a wall that excludes imagination. It is a line that keeps imagination honest. Quantum biology becomes compelling precisely when it resists the urge to inflate microscopic strangeness into cosmic significance. The cell does not levitate above thermodynamics. It pays its bills in free energy. It does not maintain coherence by spiritual calm. It does so, if at all, by molecular architecture, timescale matching, and interactions tuned by evolution.

A phrase sometimes attributed to Richard Feynman is that nature is not classical, dammit, and if you want to make a simulation of nature, you'd better make it quantum mechanical. The phrase is funny because it sounds like a tantrum, but it hides a discipline: to describe the molecular world, you must use quantum mechanics, and you must use it carefully. In biology, that care begins with the willingness to specify what is quantum, where it is quantum, how long it stays quantum, and how one would know if it were not. Anything less is not an explanation; it is a spell.

When a scientific claim survives those demands, it becomes strangely modest and strangely thrilling. Modest, because it is local and constrained. Thrilling, because it shows that life can operate in the narrow band where quantum rules are not merely the background of chemistry but an active part of function, a brief, precise ordering that holds its shape just long enough to matter.

Chapter 2

The Impossible Toolkit

Imagine a craftsman who can hammer two nails simultaneously or reach through a locked door to retrieve a tool. This chapter abandons the familiar laws of Newtonian mechanics to introduce the rule-breaking properties of the quantum realm, establishing the specific physical phenomena-superposition, tunneling, and coherence-that life has inexplicably mastered to survive.

2.1 The Art of Being Everywhere

A photon leaves the Sun and arrives, eight minutes later, on a leaf. It is tempting to picture a tiny pellet of light striking a chlorophyll molecule the way a billiard ball taps another: point-to-point, here-to-there, a neat story of contact. Yet some of the most stubborn experimental facts in physics refuse to cooperate with that picture. Light sometimes behaves like a stream of particles and sometimes like a wave that spreads out, overlaps with itself, and interferes. Matter does it too. Electrons, protons, even entire atoms can play this double role, and the rules that make it possible form the first piece of the "impossible toolkit" that quantum biology borrows: *superposition*, the ability of a quantum system to occupy multiple possibilities at once, not as a metaphor but as a mathematical and experimentally testable statement about how nature evolves.

The quickest route into the idea is the famous double-slit experiment, because it is so simple that it feels like a prank. Imagine a barrier with two narrow openings and, behind it, a screen that records impacts. If you fire paintballs at the barrier, each paintball goes through one slit or the other and lands somewhere on the screen. The final pattern is just the sum of two single-slit patterns: two blurry stripes. If you replace paintballs with waves on water, the story changes. A wave passes through both slits; each slit acts like a new source; the two wavefronts spread and overlap; and where a crest meets a crest the waves reinforce, while crest meeting trough cancels. The screen (or the water surface) ends up with an interference pattern: alternating bright and dark bands. So far, nothing mystical. Waves interfere; particles do not.

Now dim the light so far that photons arrive one at a time. Each photon lands as a single localized dot, as if it were a particle. But after many photons have accumulated, the overall distribution of dots forms an interference pattern anyway, as if each photon were a wave that went through both slits and interfered with itself. The unnerving part is not the pattern; it is the combination of *individual hits* and *collective wave structure*. Each event looks like a particle. The statistics look like a wave. Physics can handle this by giving up a classical assumption: that the photon had a definite trajectory through one slit all along.

Louis de Broglie, in 1924, proposed a bold symmetry: if light (which everyone knew as a wave) can act like a particle, perhaps electrons (which everyone knew as particles) can act like waves. Within a few years, electron diffraction was observed, and the world gained a new rule of thumb: whenever something has a well-defined momentum, it can display wave-like behavior with a wavelength inversely related to that momentum. The wave is not a smear of material like fog; it is a *probability amplitude*, a mathematical object whose squared magnitude gives probabilities for different measurement outcomes. That is why single photons and single electrons make single dots: the wave does not spread the object into a literal puddle of stuff. It spreads the *probability* of where an outcome may be found.

Superposition is the name for what that wave describes. If a system can reach the same final outcome by two alternative histories, quantum theory does not assign separate probabilities and add them. It assigns amplitudes to the alternatives and adds those, and only at the end takes an absolute square. In symbols, if two alternatives have amplitudes a_1 and a_2, then the probability is

$$P = |a_1 + a_2|^2 = |a_1|^2 + |a_2|^2 + 2\,\Re(a_1 a_2^*).$$

The extra term, the interference term, is the whole trick. It can increase a probability (constructive interference) or reduce it, even to zero (destructive interference). That is why quantum systems can behave as if they have "tested" multiple routes at once. It is not that they are secretly choosing in advance; it is that the alternatives coexist in the mathematics until something forces a definite outcome.

There is a standard trap here: picturing superposition as indecision. That sounds like psychology. Quantum superposition is more precise and stranger. It says that, prior to measurement, the system is described by a state that includes multiple mutually exclusive possibilities, and the state evolves smoothly according to wave-like equations. The oddness enters when we ask what counts as a measurement. A detector placed at the slits can tell which path a photon took, and the interference pattern disappears, replaced by the simple sum of two stripes. Merely *having the information available* about the path, even if no human watches the readout, is enough to destroy interference. Interference demands that the alternatives remain indistinguishable. In practice, the environment is an eager gossip. If the world around a particle can learn which route it took, the delicate phase relationship between alternatives leaks away. The resulting loss of phase coherence is called *decoherence*, and it is the reason quantum behavior is so hard to maintain in big, warm, complicated objects.

Yet "hard" is not the same as "impossible," especially on the molecular scale. Biology lives there: not at the size of billiard balls, but at the size where electrons and excitations are naturally quantum objects, and where the environment is not merely a nuisance but a structured, dynamic partner. The first question for quantum biology is not whether quantum mechanics applies (it always does), but whether superposition and interference survive long enough, and in the right degrees of freedom, to matter for function rather than being washed out into ordinary chemistry.

A good way to feel the difference between classical and quantum exploration is to imagine a maze. A classical mouse is a heroic little algorithm. It runs a corridor, hits a dead end, turns around, tries another branch, and gradually learns. Even if the mouse is very fast, it still has to pay the price of serial search: one path after another. A quantum system, by contrast, can spread into a superposition over many paths. Its amplitude flows down every corridor at once, and when those corridors later reconnect, the amplitudes interfere. In a carefully designed network, wrong turns can cancel themselves out while right turns reinforce. The mathematics resembles a *quantum walk*, the quantum cousin of a random walk, and it is one of the most useful mental models for why "being everywhere" might provide an advantage.

Place the following diagram immediately after this point: it makes the maze metaphor concrete without pretending that molecules are literally mice.

Classical mouse (one route at a time)

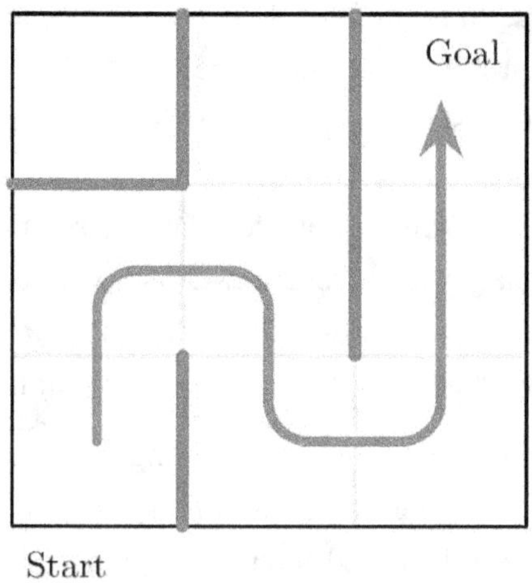

"Quantum mouse" (mist through all routes)

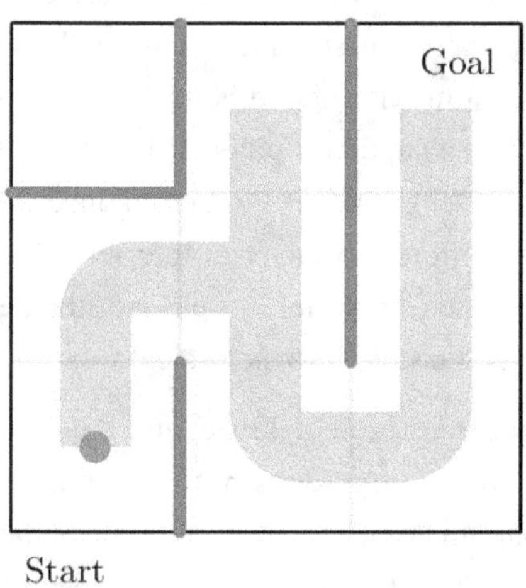

If the maze metaphor feels fanciful, it becomes less so when translated into molecular terms. Photosynthesis, after all, is a problem of routing. A pigment absorbs a photon and becomes electronically excited. That excitation is not a physical particle drifting through space; it is a redistribution of electronic energy that can be shared among pigments. In many light-harvesting complexes, pigments sit close enough that an excitation can be *delocalized* across several of them: the quantum state is a superposition in which “the excitation is on pigment A” and “the excitation is on pigment B” are both part of the description. From the outside, energy seems to migrate with remarkable efficiency to a reaction center where it can be stored chemically. The biological question is brutally practical: how does the excitation avoid getting stuck, wasted, or sent the wrong way?

Classically, one might imagine a hopping process. The excitation sits on one pigment, then jumps to a neighbor with some rate, then jumps again, like a drunkard navigating stepping stones. This can work, and often does. But in a disordered landscape, hopping has a weakness: it can become trapped. Local energetic mismatches can create dead ends where it is statistically more likely to hop back than forward. The quantum alternative is not that the excitation literally tries every path like a supercomputer. It is that the excitation can be in a superposition over multiple pigments at once, and the amplitudes for moving along different routes can interfere. Even modest coherence can allow a kind of parallel sampling of the network before the environment turns the state into a classical mixture.

Here the story becomes more delicate, and it is important not to sell miracles. Quantum biology does not require a leaf to behave like a laboratory quantum computer, keeping immaculate superpositions for milliseconds. The relevant timescales in electronic energy transfer can be femtoseconds to picoseconds. A femtosecond is 10^{-15} seconds, a timescale so short that even molecular vibrations barely get a chance to move. Coherence that persists for a few hundred femtoseconds can still influence which route an excitation tends to take, simply because those early instants are when the wave-like spreading and interference are strongest. That is why experiments claiming coherence on such timescales triggered so much excitement: they suggested that quantum interference might be present long enough to shape biological outcomes.

The most celebrated case involved a photosynthetic complex called FMO (Fenna—Matthews—Olson), found in green sulfur bacteria. Around 2007, Greg Engel and colleagues used two-dimensional electronic spectroscopy, a technique that can resolve ultrafast dynamics with a kind of temporal and spectral "fingerprint," and reported oscillatory signals that were widely interpreted as evidence of quantum coherence among excitonic states. A few years later, work led by Gitt Panitchayangkoon and collaborators extended such observations toward physiological temperatures and reported coherence persisting on the order of hundreds of femtoseconds. The narrative that quickly formed was irresistible: perhaps nature discovered a way to keep quantum superpositions alive in a warm, wet protein scaffold, using structure and motion to protect the phase relationships that allow interference.

Science, being less romantic than headlines, did not stop there. The same experimental signatures can sometimes be produced by different underlying mechanisms. Oscillations in spectroscopic data might reflect electronic coherence, or vibrational coherence, or a hybrid known as vibronic coherence; they might also be sensitive to how signals are processed and interpreted. In 2016, Hong-Guang Duan, Valentyn Prokhorenko and colleagues argued that electronic coherence in such systems decays much faster at ambient conditions, closer to tens of femtoseconds, and that longer-lived oscillations need not indicate long-lived electronic superposition. The dispute is technical, but the broader lesson is human and familiar: nature often offers ambiguous clues, and deciding which interpretation is correct demands both better experiments and better theories.

Still, even if electronic coherence is short-lived, superposition need not be irrelevant. There is a quieter, more robust idea that emerged from theoretical work by Masoud Mohseni, Patrick Rebentrost, Alán Aspuru-Guzik and collaborators around 2008—2009: *environment-assisted quantum transport*. The phrase sounds like a contradiction, since the environment is usually the villain that causes decoherence. But a moderate amount of interaction with the environment can sometimes help, not hurt. The reason is that perfectly coherent evolution in a disordered network can lead to *localization*, where wave interference traps amplitude in certain regions. A little dephasing noise can break the stalemate, allowing the excitation to escape traps and find the reaction center more reliably. Too much noise, and the system becomes purely classical hopping again. Too little, and it may remain stuck. Between them lies a "Goldilocks" regime where some coherence exists but does not remain pristine. Gregory Scholes, Graham Fleming and others later emphasized that functional coherence does not necessarily mean long-lived coherence; it means coherence that persists long enough to influence a decision the system makes, even if that decision is made quickly.

It is worth pausing over what superposition really buys you, because "being everywhere" can sound like cheating. The advantage is not infinite speed. A quantum excitation in a light-harvesting complex is not solving a maze designed by an adversary. It is moving in a landscape sculpted by evolution, where pigment energies, couplings, and protein motions have been tuned within chemical constraints. In that setting, a short burst of wave-like spreading can be a way of *not committing too early*. If multiple routes lead to the reaction center, the excitation can distribute amplitude among them. If one route is blocked by an energetic mismatch, interference and environmental interactions can reduce its contribution while favoring others. The system does not need to "know" which route is best; it needs dynamics that tend, on average, to deliver energy efficiently before it is lost.

Wave—particle duality also reshapes the meaning of a "path." In classical mechanics, a path is a continuous line traced by an object. In quantum mechanics, a path is one of the alternatives included in the superposition, and the object is not required to occupy a single one until measurement-like interactions force it to. Richard Feynman captured this in his path integral formulation, where the amplitude for a particle to go from one place to another is built from contributions of many conceivable paths, each weighted by a phase. While that formalism is often taught with a flourish, its moral is practical: quantum systems are sensitive to the *relative phases* between alternatives. Change the environment, the energy landscape, or the timing, and you change the interference pattern—sometimes subtly, sometimes dramatically.

In a biological setting, phases are not abstract numbers floating in space; they are tied to energies, couplings, and motions. Every pigment sits in a protein environment that jiggles thermally. Those motions shift energies and scramble phases, but they can also create correlations: neighboring pigments may experience related fluctuations, and such shared noise can preserve certain phase relationships longer than uncorrelated noise would. The protein is not just scaffolding. It is part of the dynamical system that determines which superpositions are possible and how quickly they degrade. That is one reason the question "Is photosynthesis quantum?" is too blunt. The more precise question is: *Which degrees of freedom remain coherent, for how long, under which conditions, and with what functional effect?*

A concrete example helps. Imagine two pigments, A and B, close enough that the excitation can be shared. If the excitation is localized on A, that is one state, call it $|A\rangle$. Localized on B is another, $|B\rangle$. If the coupling between them is strong compared with the environmental noise on the relevant timescale, the natural eigenstates are not $|A\rangle$

and $|B\rangle$ but superpositions like

$$|\pm\rangle = \frac{1}{\sqrt{2}} \left(|A\rangle \pm |B\rangle\right),$$

with slightly different energies. In such a pair, the excitation can oscillate between A and B, and interference effects can determine how readily it transfers onward to a third site, C. If site C couples more strongly to the $|+\rangle$ combination than to the $|-\rangle$ combination, then the phase relationship between A and B matters. Decoherence that randomizes the relative phase between A and B effectively turns the superposition into a classical mixture, reducing the selectivity. Even a brief window of coherence can bias the transfer.

If this feels like an obscure detail, it has a familiar analog in everyday wave physics. In a concert hall, two speakers playing the same tone can create spots where the sound is louder and spots where it nearly disappears. The air is noisy and warm, yet interference is still real because the wavelength is large and the phase relationship is maintained. In molecules, the "speakers" are quantum states, and the "air" is the protein and solvent. The challenge is that molecular environments are efficient at disrupting phase. Biology cannot change the basic fact that warm environments decohere. What it can do is shape the couplings and timescales so that whatever coherence exists is not wasted.

There is also a psychological advantage to taking superposition seriously: it prevents a common misunderstanding that quantum biology must be exotic. At the molecular scale, quantum states are the natural language. Electron transfer, bond formation, and optical absorption already rely on quantum mechanics. The controversial step is not invoking quantum theory, but invoking *non-trivial* quantum features such as interference between alternative pathways that persist long enough to be functionally consequential. Superposition is therefore best treated not as magic dust sprinkled on biology, but as a specific mechanism with clear conditions for when it matters: alternative histories must remain coherent and indistinguishable long enough for their amplitudes to interfere.

A last historical note brings the strangeness back down to earth. When Niels Bohr and his contemporaries argued over what quantum theory meant, the debate sometimes sounded like metaphysics. But the double-slit experiment is stubbornly operational. Interference appears or disappears depending on what information is available about the path. That practical attitude is also the healthiest way to approach quantum biology. Ask not whether a living system is "in superposition" in some grand philosophical sense, but whether there are experimentally accessible signatures of phase coherence and interference in the relevant molecular degrees of freedom, and whether manipulating those signatures changes function.

The leaf does not need to keep a photon in a mysterious limbo. It needs to move energy quickly, amid disorder, before the excitation dies. If the excitation can briefly behave like a wave—spreading, sampling, and interfering across a network of pigments—then the leaf gains a subtle edge: not omniscience, not perfection, but a kind of early-time flexibility. In a world where every absorbed photon is a chance to store energy or to lose it as heat, even that modest advantage can be worth having.

2.2 Walking Through Walls

Stand with a tennis ball at the foot of a hill and the rules feel obvious. If you do not throw hard enough, the ball rolls partway up, slows, and returns to your feet. That simple fact is so dependable that much of engineering is just its careful bookkeeping: how much energy goes in, how much friction steals, how much remains to crest the barrier. Chemistry is full of hills too. Molecules do not transform smoothly from reactants into products; they must usually climb an "activation barrier," a transient, high-energy arrangement of atoms where old bonds are strained and new ones are not yet secure. At room temperature, thermal jostling can sometimes supply the push. When it cannot, reactions crawl.

Quantum mechanics changes the story in a way that still sounds like a conjuring trick when phrased badly. A particle facing an energy barrier higher than its available energy can nevertheless be found on the far side. It has not borrowed energy, or drilled a hole, or squeezed through a microscopic crack. Instead the mathematics that describes it extends into the forbidden region, and with some probability it appears beyond. The name for this is *quantum tunneling*. The name is good: it suggests a route through an obstacle, but it is also misleading, because it tempts the mind to imagine a tiny bullet carving a tiny tunnel. Nothing of the sort happens. What happens is that the particle is described by a wave-like state, and wave-like states do not stop sharply at the place where a classical particle would be turned back. They fade.

A minimal picture is enough. In classical physics, a particle has a definite position and a definite energy. Where the barrier exceeds that energy, the particle simply cannot go. In quantum physics, a particle is described by a wavefunction, a probability amplitude whose square gives the chance of finding the particle in a given place. When this wave encounters a barrier, part of it reflects, as a water wave reflects from a wall, and part penetrates. Inside the barrier the amplitude does not oscillate as an ordinary traveling wave; it decays exponentially. If the barrier is thin enough, a small remnant of amplitude reaches the far side and becomes a traveling wave again. When a detector checks, it either finds the particle reflected or transmitted. No intermediate half-particle is seen; the strangeness lies in the probabilities.

The exponential nature of that decay is the main reason tunneling is both ubiquitous at the molecular scale and rare in daily life. The transmission probability depends roughly on three things: how high the barrier is, how wide it is, and how heavy the particle is. Make the barrier slightly wider and tunneling plummets. Replace an electron with a proton and tunneling drops dramatically. Replace a proton with a tennis ball and the probability becomes so absurdly small that one may as well call it zero. Tunneling is not a loophole in energy conservation; it is a fragile consequence of wave mechanics that is easily suppressed by mass and distance.

The historical route to tunneling is one of those episodes where an apparent detail becomes a profound clue. In 1928, George Gamow used tunneling to explain alpha decay in nuclei. Classically, an alpha particle trapped inside a nucleus should never escape: the nuclear forces hold it in and an electrostatic barrier blocks the way out. Yet certain nuclei emit alpha particles at well-defined rates. Gamow showed that if the alpha particle has a wavefunction that leaks through the barrier, the observed decay rates can be understood as tunneling probabilities. The astonishing thing was not just that tunneling existed, but that it could be used to compute measurable lifetimes, turning a paradox into a quantitative success.

For chemistry, tunneling has long been part of the underlying quantum machinery. Electrons tunnel routinely; without it there would be no modern electronics, no scanning tunneling microscope, and much of chemical bonding would make little sense. The more biologically provocative question concerns heavier nuclei, especially hydrogen. A hydrogen nucleus is a proton (or, in deuterium, a proton plus a neutron), and because it is the lightest nucleus, it has the best chance of tunneling among nuclei commonly moved in biochemical reactions. Biology is full of hydrogen transfers: enzymes shuttle hydrogens during metabolism, DNA bases rearrange hydrogens during rare tautomeric shifts, and proteins shift protons to control charge and shape. If tunneling contributes meaningfully to any of these steps, then an enzyme is not merely lowering a classical barrier; it is also shaping a quantum probability.

A good way to keep tunneling from turning into mysticism is to focus on the barrier as a *landscape* rather than a wall. Chemical reactions can be drawn as a curve of potential energy versus a reaction coordinate, a collective measure of how far the system has progressed from reactants to products. The activation barrier is the hilltop. Classically, the system must climb to the top. Quantum mechanically, the system has a nonzero amplitude to move from one side to the other without ever occupying the hilltop in the classical sense. The barrier is not bypassed by speed; it is bypassed by the peculiar rule that probability amplitudes can extend where classical motion cannot.

That still leaves the most interesting part: what does an enzyme actually do? Popular descriptions often say that an enzyme "lowers the activation energy." That is true but incomplete, like saying a violin makes sound by vibrating. Enzymes are not rigid jigs. They are soft, fluctuating structures in water, and their active sites are crowded with charges and polar groups that tug, twist, and stabilize different atomic arrangements. Those motions are often dismissed as "thermal noise," yet for tunneling they are central. Tunneling is exquisitely sensitive to distance. For a hydrogen to tunnel from a donor–acceptor atom to an acceptor atom, the donor–acceptor separation must be short enough and the electrostatic environment must be favorable enough for the wavefunction to have appreciable overlap. Proteins breathe. Most conformations are not optimally aligned. Some fraction of the time, thermal motion brings the donor and acceptor into a "tunneling-ready" geometry, and then the transfer becomes vastly more probable.

Judith Klinman and Amnon Kohen, working through the 1990s and into the 2000s, helped make this picture experimentally persuasive by treating tunneling not as an exotic add-on but as something one can diagnose with careful kinetic measurements. One of the most informative tools is the *kinetic isotope effect*. Replace a transferring hydrogen (protium) with deuterium, and if the reaction were purely classical, the rate would change in a fairly predictable way, largely through differences in zero-point energies and vibrational frequencies. When tunneling plays a significant role, isotope effects can become unusually large, and their temperature dependence can deviate from classical expectations. The logic is almost embarrassingly concrete: if a reaction slows down far more than it "should" when the hydrogen is made heavier, and if the way it slows changes with temperature in a distinctive pattern, then a quantum contribution to nuclear motion is hard to avoid.

This does not mean enzymes are teleporting atoms. Tunneling still obeys the same iron laws of sensitivity to distance and barrier shape. What enzymes appear to do is tune those quantities. The active site can position the donor and acceptor close together, polarize bonds to thin the barrier, and organize an electrostatic field that makes the product side energetically attractive. At the same time, the protein and surrounding water reorganize as charge moves, a process often described as *reorganization energy*. Hydrogen transfer in enzymes frequently occurs as part of a coupled motion of electrons and protons, so-called proton-coupled electron transfer. Sharon Hammes-Schiffer has been influential in emphasizing that one must treat the proton quantum mechanically while also accounting for the shifting environment. The proton does not tunnel through a fixed barrier drawn on a chalkboard. It tunnels through a barrier that is itself being reshaped by molecular motion, which can help or hinder the transfer.

A kitchen-table analogy helps, as long as it is handled gently. Imagine trying to pass a note from one person to another across a crowded room. If everyone holds perfectly still, the straight path may be blocked. But if the crowd sways, gaps open momentarily. You do not push through the wall of people; you wait for a fluctuation that creates a passage. For enzymes, thermal motion creates fleeting alignments where the effective barrier for hydrogen transfer becomes thin and well-coupled. Tunneling is then the quantum part: the probability that, in that fleeting alignment, the hydrogen appears on the other side without climbing the full classical hill.

The word "appears" deserves respect. It is easy to imagine tunneling as a particle burrowing under the hill, tracing a hidden trajectory. Quantum mechanics does not require such a picture. The proton is described by a spread-out wavefunction in the direction of transfer, and the transfer event is a change in where the wavefunction is localized when the system settles into one of the stable configurations. In many enzyme reactions, the relevant coordinate is not the proton alone but a collective coordinate involving bond vibrations and environmental polarization. The proton's quantum character is entangled, in a loose chemical sense, with the motion of heavier atoms that modulate the barrier. One can still speak of tunneling as a distinct mechanism, but it is not separable from its context.

A concrete example often discussed in the literature is alcohol dehydrogenase, an enzyme that transfers a hydride (a hydrogen nucleus plus two electrons) or, in related reactions, a hydrogen as part of redox chemistry. Experiments measuring rates with isotopic substitution have produced signatures consistent with substantial quantum contributions. The details vary by enzyme and reaction, and no honest account should claim tunneling is always dominant. The deeper point is methodological: the question is no longer "Can tunneling happen in biology?" but "When does it contribute appreciably, and how does the protein scaffold modulate it?" That shift turns tunneling from a parlor trick into an ordinary scientific variable, something to be measured, modeled, and, potentially, designed around.

Design is where the story becomes emotionally resonant, because it touches medicine. If an enzyme's catalytic power partly comes from arranging a tunneling-ready geometry, then an inhibitor that subtly perturbs donor–acceptor distances or active-site dynamics could be effective even if it does not look dramatic in a static structure. Drug discovery has long relied on the lock-and-key metaphor: find a molecule that fits the active site and blocks it. But proteins are not locks; they are restless. A tunneling-sensitive step makes that restlessness part of the mechanism. In that view, an effective inhibitor might work not by blocking a path but by making the right fluctuation rarer, nudging the active site away from the geometry that permits efficient hydrogen transfer. The practical consequence is not that medicinal chemists must solve the Schrödinger equation for every drug candidate. It is that mechanistic understanding can widen the set of plausible strategies: geometry, electrostatics, and dynamics can matter in concert, and small changes can have large effects when tunneling probabilities are exponential in distance.

Tunneling also appears in a more speculative and culturally charged context: mutation. DNA bases can exist in different tautomeric forms, rare rearrangements where a proton shifts position and hydrogen-bonding patterns change. In principle, such a shift could lead a base to pair incorrectly during replication, introducing an error. It is an old idea, often associated with early speculations by Erwin Schrödinger in *What is Life?* and with later discussions in molecular biology, that quantum effects might contribute to mutation rates. The difficulty is that DNA in cells is bathed in water, buffeted by proteins, and held in structures that strongly constrain and stabilize particular forms. The mere fact that a proton can tunnel in principle does not mean it does so often enough to matter biologically.

Recent theoretical work has sharpened the debate rather than settling it. Some models of proton transfer in adenine–thymine base pairs suggest that the probability of forming the rare tautomer via tunneling under physiological conditions is extraordinarily small, implying that tunneling is unlikely to be a major direct source of mutations in that context. Other open-system models, focusing on guanine–cytosine and including different assumptions about environmental coupling and barrier shapes, have argued for larger tunneling contributions and higher transient tautomer populations. Both can be mathematically respectable while disagreeing, because tunneling is hypersensitive to the details of the potential energy surface and to how the environment is represented. That sensitivity is not a flaw; it is the mechanism. It also means that deciding what is biologically relevant demands more than plausibility arguments. It demands constraints from experiment: isotope substitution studies in relevant contexts, spectroscopic probes where possible, and careful links between microscopic events and macroscopic mutation rates.

The temptation, when faced with such uncertainty, is to retreat to safe statements: "Quantum mechanics underlies all chemistry." True, but unhelpful. The more interesting stance is narrower and sharper. Tunneling is not a magical shortcut; it is a probability that can be enhanced or suppressed by structural details. Biology is a master of structural details. Proteins can hold atoms at just-so distances and charge distributions, and they can choreograph motions on timescales that match chemical steps. When that choreography brings light nuclei close to their targets across thin barriers, tunneling becomes an efficient route. When it does not, tunneling becomes negligible. Warmth and wetness do not banish tunneling; they simply make it contingent.

One last way to keep intuition honest is to remember that "through a wall" is not always the right image. In many biochemical hydrogen transfers, the barrier is not a fortress but a ridge, and the reactant state and product state are two valleys separated by a saddle. Thermal motion can lift the system partway up the ridge, thinning what remains. Tunneling then handles the last part, the classically forbidden remainder. In that blended picture, classical and quantum effects cooperate. The protein's thermal dance positions and polarizes, reducing and reshaping the barrier, and the hydrogen's quantum nature takes advantage of the moment. The result is not a violation of everyday logic, but an extension of it: not everything that crosses a barrier has to climb to the highest point.

For a living cell, the payoff of such subtlety is measured in rates and reliability. A metabolic pathway is a relay race where the baton is passed as quickly as possible before it is dropped. If a key step is slowed by a barrier, the whole pathway bottlenecks, with consequences for growth, stress responses, and survival. Evolution does not need to "understand" tunneling to exploit it. It only needs variants whose active sites, by chance, make favorable geometries more common, or make barriers thinner, or align electrostatics in ways that increase transfer probabilities. Over deep time, tiny advantages can be sculpted into robust mechanisms. A proton does not need to walk over the hill every time; sometimes, in the right molecular posture, it can simply be found on the other side.

2.3 The Rhythm of Reality

A wave is not just a shape in space. It is also a timing. Two people can clap at the same volume, yet if they drift out of sync the room loses its sharp, collective beat. Quantum effects that rely on interference have the same vulnerability. Superposition allows multiple alternatives to coexist as probability amplitudes; interference decides whether those alternatives reinforce or cancel. The ingredient that makes interference possible is *coherence*: a stable relationship between the phases of those amplitudes, a kind of internal synchronization. Lose that phase relationship and the system still obeys quantum mechanics, but it begins to *behave* classically because the telltale interference patterns wash out.

Coherence is easy to describe and notoriously hard to keep. In a physics lab, coherence is protected by cold temperatures, vacuum chambers, electromagnetic shielding, and exquisite isolation. Life, by contrast, happens in warm water, packed with moving molecules, vibrating bonds, fluctuating electric fields, and constant collisions. The central drama of quantum biology lies here. Wave-like behavior and tunneling are not exotic in themselves; the mystery is whether the coherence needed to make them functionally significant can survive long enough to matter, and whether biology can tune the inevitable noise into something less like vandalism and more like a rhythm section.

Coherence is not a mystical glow. It is a property of a quantum state. Consider again the simple case of an excitation shared between two sites, described by a superposition $|\psi\rangle = \alpha|A\rangle + \beta|B\rangle$. The complex numbers α and β carry both magnitudes (how much weight each alternative has) and phases (their timing relative to each other). If the relative phase between α and β is well-defined, the system is coherent and interference can occur. If the environment scrambles that relative phase unpredictably, then even if the magnitudes stay the same, interference disappears. Mathematically, one can describe this loss by saying that in the reduced description of the system (ignoring the countless environmental degrees of freedom), the off-diagonal terms of the density matrix decay. In plain language: the environment steals the phase information and hides it in correlations too complex to track.

This process is called *decoherence*. It is not primarily about energy loss. A system can lose coherence without losing much energy at all. It is about information: information about *which alternative happened*. The double-slit experiment makes this painfully clear. When no record exists of which slit a particle used, the alternatives remain indistinguishable and can interfere. When the environment can, even in principle, carry away a trace that distinguishes the alternatives, interference vanishes. In biology, there are countless ways to leave such traces: a nearby dipole rotates slightly differently depending on where an electron sits, a water molecule shifts its hydrogen-bonding, a protein side chain jiggles. Any of these can act like a witness that ruins the secret needed for interference.

At this point a discouraging conclusion seems unavoidable: coherence should die almost instantly in living systems, so quantum biological effects beyond trivial chemistry should be negligible. Many scientists have said versions of this, and it remains a healthy skepticism. Yet there is a countervailing fact that makes the question worth asking rather than dismissing. Biology is not a random warm soup. It is structured matter. Proteins create pockets that exclude water, align charges, and constrain motion. Pigments are held at specific separations and orientations. Reaction centers are engineered by evolution to accept energy and charge in particular ways. Structure can slow down some forms of decoherence and can create regimes where coherence persists briefly but long enough to influence an outcome.

Briefly can be enough. In the context of photosynthetic energy transfer, a few hundred femtoseconds is already a relevant timescale. In the context of electron and spin dynamics, microseconds can be relevant. There is no single required coherence time; there is only the demand that coherence persist for the time it takes to do something useful. A tuning fork does not need to ring for an hour to set a rhythm. It needs to ring long enough for the next instrument to lock on.

A historical anecdote makes the fragility of coherence less abstract. When Erwin Schrödinger wrote down his wave equation in 1926, it was natural to treat the wavefunction as something physically real, perhaps like a smeared-out charge cloud. Within a decade, the uneasy marriage between wave-like evolution and particle-like outcomes forced a different view. Schrödinger himself proposed a famous thought experiment in 1935, a cat entangled with a quantum event in a sealed box, to dramatize how absurd it would be to treat macroscopic objects as maintaining coherent superpositions indefinitely. His point was not that cats are spooky. His point was that the environment of a macroscopic object acts like a measurement apparatus, destroying coherence so rapidly that superpositions of distinct macroscopic states do not persist in any practical sense. Today that intuition is formalized in decoherence theory. Macroscopic objects are not exempt from quantum laws; they are simply too entangled with their environments for coherent alternatives to remain accessible.

That word, *entangled*, is the second tool in this part of the kit, and it is more easily abused than any other. Entanglement is a particular kind of quantum correlation between parts of a system, one that cannot be explained by assigning each part its own independent state. If two coins are classically correlated, knowing one tells you something about the other, but each coin still has a definite face, even if you do not know it. Entangled systems are different. A pair can be in a joint state where neither part has a definite state on its own, yet the pair has well-defined joint properties. Measure one and you instantly update what you can predict about the other. That instantaneous update is what Einstein derided as "spooky action at a distance," though the spookiness is largely a matter of language: no usable signal travels faster than light, but the correlations are real and have been tested repeatedly.

Entanglement is not automatically useful in biology, and invoking it does not automatically make a claim deeper. But it provides a sharp way to talk about how a system can behave as a coordinated whole, even when its parts are separated. It also provides a diagnostic challenge. Entanglement is fragile, typically even more fragile than simple coherence, because it requires maintaining coherent phase relationships across multiple degrees of freedom. In a warm environment, entanglement can vanish quickly. That is one reason entanglement claims in biology must be treated with care. The most plausible cases tend to involve degrees of freedom that couple weakly to the environment or have naturally long coherence times.

This brings us to a biological phenomenon that has captured imaginations for decades: animals navigating using Earth's magnetic field. Migratory birds, sea turtles, some fish, and even insects appear to extract directional information from the geomagnetic field, which is weak—tens of microtesla—and easily overwhelmed by everyday magnetic clutter. How could a biological sensor possibly respond to such a faint cue?

One leading explanation is the *radical pair mechanism*, proposed in the 1970s and developed substantially by many researchers since. The basic ingredients are chemical and familiar. Light (often blue light) excites a molecule and triggers an electron transfer, creating two radicals: molecules or molecular fragments with unpaired electron spins. These two unpaired spins begin in a correlated quantum state, typically a *singlet* state, where the spins are paired in opposite directions in a way that makes the total spin zero. Under the influence of internal magnetic interactions (notably hyperfine couplings to nearby nuclear spins) and the external geomagnetic field, the pair can oscillate between singlet and *triplet* character, where the total spin is one and the spins are aligned in a correlated way. Crucially, singlet and triplet radical pairs can have different chemical fates: they may recombine at different rates or produce different products. A weak magnetic field can therefore change the yield of reaction products by subtly changing the spin dynamics.

The remarkable feature is that this mechanism depends on *spin coherence*. The singlet—triplet interconversion is a quantum interference phenomenon in spin space. If spin coherence is lost too quickly, the oscillations wash out and the chemical yields become insensitive to the magnetic field. If coherence persists long enough relative to reaction timescales, the yields can depend on field strength and direction. The animal then needs only to connect those chemical yield differences to a neural signal, perhaps via changes in signaling molecules or receptor states.

The best-known molecular candidate for hosting such radical pairs is *cryptochrome*, a light-sensitive flavoprotein found in plants and animals and implicated in circadian rhythms. The appeal of cryptochrome is that it is present in retinas, it can undergo photoinduced electron transfer, and it offers a plausible biochemical stage for radical pair formation. None of this is a final proof that cryptochrome is *the* magnetic compass in all species, and multiple mechanisms may coexist. But it provides a concrete molecular anchor for an otherwise ghostly behavioral phenomenon.

Behavioral evidence has sharpened the intrigue. Experiments have reported that certain birds become disoriented under specific radiofrequency (RF) noise conditions, consistent with the idea that a spin-dependent mechanism is being perturbed. The pattern of light dependence in some species also aligns with a photochemical compass rather than a magnetite-based physical compass. These findings are not universally straightforward to interpret—behavior is noisy, animals are complicated, and experimental conditions matter—but they help explain why magnetoreception became a flagship topic for quantum biology. It offers a clear link between a quantum-sensitive degree of freedom (electron spin) and a macroscopic function (navigation).

Here is where entanglement enters with both promise and peril. A radical pair created from a single electron transfer can begin in an entangled spin state. In fact, the singlet state is an archetype of entanglement. If that entanglement persists for a biologically relevant time, it could, in principle, enhance sensitivity or enable richer directional information. Some theoretical analyses have explored whether entanglement measures correlate with compass performance. Yet it is easy to overstate what is required. The radical pair mechanism can be genuinely quantum even if entanglement is fleeting, because the key ingredient for magnetic sensitivity is coherent singlet—triplet mixing, not necessarily long-lived entanglement by any strict technical measure. A useful discipline here is to ask what the system *needs* to function: it needs phase-sensitive spin dynamics on the timescale before recombination. Whether that phase sensitivity is best described as entanglement at each moment is a secondary, and sometimes misleading, question.

Modeling has also complicated the picture in an instructive way. C. T. E. Rodgers and P. J. Hore, among others, have emphasized that semiclassical approximations—treating some spins classically or averaging over quantum phases—can fail in regimes likely relevant to realistic radical pairs, particularly when electron-spin coupling is present. That is a technical statement with a philosophical bite. It suggests that "quantum" here does not mean merely that electrons obey quantum mechanics in principle; it means that to predict the compass signal correctly, one may need an explicitly quantum treatment of the spin dynamics. The requirement is not aesthetic. It is forced by the mathematics of interference and coherence.

Noise, again, plays an ambiguous role. In radical pair systems, certain kinds of environmental interactions can both destroy and create useful effects. Too much decoherence and the compass dies. Too little interaction and other constraints can dominate, such as overly sharp resonances that are impractical in biological variability. Some noise sources can broaden responses, making the mechanism more robust to disorder. The lesson resembles the “Goldilocks” theme that appears in discussions of energy transport: living systems may benefit not from eliminating noise, but from existing in a regime where coherence is present yet not pristinely isolated. Biology does not build vacuum chambers. It builds structured, imperfect shelters.

Coherence has a second, less glamorous but equally important meaning in this context: it is a bookkeeping device for what the system remembers about itself. A coherent quantum state carries phase information that encodes the history of its evolution. Decoherence is forgetting. A protein environment that scrambles phases is a crowd that drowns out the beat; a structured environment that preserves certain phase relationships is a concert hall that supports resonance. This metaphor has limits, but it captures something real: quantum coherence is relational. It is not a substance the system possesses; it is a pattern of correlations that can be sustained or destroyed depending on the surrounding dynamics.

One might ask, reasonably, why nature would ever rely on something so fragile. The honest answer is that it may not rely on it in the strong sense. Many biological processes work perfectly well in classical or semiclassical regimes. The more plausible claim is narrower: when the relevant degrees of freedom are already quantum (electronic excitations, electron spins, hydrogen nuclei), and when the timescales of function are short enough or the coupling to the environment is weak enough, coherence can persist long enough to bias outcomes. That bias can then be amplified by ordinary biochemical feedback into a macroscopic effect.

A final image helps keep coherence and entanglement grounded. Imagine a pair of ice skaters performing a routine. If they are linked by a firm grip and share timing, their motions are coordinated in a way that cannot be reduced to two independent skaters doing their own thing. That coordination is not supernatural; it is a real constraint on the combined system. Now imagine the rink becomes rough and crowded. Bumps and distractions force them to loosen their grip, their timing slips, and soon they are simply two skaters trying not to fall. In quantum mechanics, coherence is the shared timing, and entanglement is the grip that makes the pair a single choreographed unit. Decoherence is the crowd.

In living matter, the crowd is always present. The extraordinary question is not whether the crowd exists, but whether, in carefully chosen corners of biology, the choreography can remain intact long enough to do useful work. If a bird can read direction from a whisper of magnetism because a fleeting entangled radical pair keeps time for microseconds, or if a protein can steer a reaction because a hydrogen wavefunction retains its phase long enough to tunnel efficiently, then the "impossible" part is not that quantum mechanics intrudes into life. It is that life, without ever trying to be quantum, has found ways to make just enough room for the rhythm to hold.

Chapter 3

Drinking the Light

Life depends on capturing the sun's fleeting energy, yet the classical laws of physics suggest most of this energy should be lost as heat before it becomes useful fuel. We dive into the molecular machinery of photosynthesis to find an architecture so efficient it suggests biology has tapped into the quantum realm to survive.

3.1 Architecture of a Sun Trap

A leaf looks like a flat green bargain with the Sun: a thin sheet of matter that somehow turns light into the substance of living things. The trick sounds straightforward until you take the leaf seriously as an engineering object. Sunlight arrives as individual photons, scattered across a surface by the wind, clouds, and the daily sweep of the Sun. Each photon carries a fixed packet of energy, but it arrives at an essentially random location. The chemistry that can *store* that energy, however, is not spread uniformly across the leaf. It is concentrated in specialized molecular machines embedded in membranes, where an absorbed photon can be converted into a stable separation of electrical charge—the first down payment on sugars, starch, and the rest of biomass.

Between those two facts lies the central architectural problem: light is caught over a wide area, while the "battery" is small, buried, and picky about what it accepts. On the scale of atoms, the separation between "where a photon hits" and "where its value is banked" is enormous. A few nanometres is the width of a protein; tens of nanometres is already a small neighbourhood in molecular terms. Yet the energy must move quickly enough that it does not leak away as heat or light, and reliably enough that the cell does not cook itself under bright noon Sun. Photosynthesis is often praised for its efficiency, but that efficiency is not just chemistry; it is architecture: a built environment of proteins and pigments designed to shepherd a fragile form of energy through a crowded, noisy world.

The "pigments" are the parts that actually interact with light: chlorophylls in plants and many algae; bacteriochlorophylls in certain bacteria; carotenoids that broaden absorption and provide protection. The proteins are the scaffolding: they hold pigments at particular distances and orientations, tune their energies, and couple the light-catching pieces to the charge-separating core. Together they form light-harvesting complexes, or LHCs. "Complex" is not a term of flattery. These are densely packed molecular assemblies in which the positioning of each pigment matters, because a pigment in the wrong place is not merely useless; it can become dangerous, creating reactive oxygen species that damage the cell.

If all of this sounds like technology, that is because biology has converged on the same design constraints that human solar engineers face, only with different materials and a vastly smaller budget per device. An antenna must intercept light efficiently. It must work in dim conditions, and it must also survive in bright ones. It must deliver energy to a conversion centre rapidly. And it must do this in a medium where everything is jiggling, bumping, and being chemically attacked.

One way to feel the difficulty is to compare the physical size of the target to the physical size of the capture area. A reaction centre—

the molecular site where the first stable charge separation occurs—is a protein complex only a few nanometres across. Even if there are many reaction centres spread throughout a photosynthetic membrane, they still represent sparse, discrete sinks. Meanwhile a cell's outer surface, or the membrane area exposed to light, is huge by comparison. If pigments were placed only inside reaction centres, most photons would simply miss. Plants and bacteria solve this by building a distributed antenna system: many pigments collect photons and funnel the resulting excitation energy inward toward fewer reaction centres.

That word "funnel" is more than a convenient metaphor. It is an energetic shape imposed on a physical shape. Pigments are not identical; their exact absorption energies depend on their chemical identity and on their environment. A chlorophyll nestled in one pocket of a protein does not have exactly the same energy as a chlorophyll in another pocket. Proteins exploit this sensitivity. They create a landscape in which energy tends to move downhill, from pigments that absorb higher-energy light (shorter wavelength) toward pigments tuned to slightly lower energies (longer wavelength). If energy is water, the protein arranges the channels and the slope.

This is also the point where the architecture departs from what we might naively expect. A simple funnel would just be a gradient: outer pigments higher, inner pigments lower, and the energy rolls in. But in a molecular world, rolling is not a thing. The energy is stored in electronic excitation: an electron has been promoted to a higher-energy state, leaving behind a positively charged "hole". That excited state can relax by emitting a photon (fluorescence), by transferring energy to vibrations (heat), or by handing the excitation to a neighbour pigment. The last option is the one that keeps the energy in play. Yet handing off is not guaranteed. It depends on distances, orientations, and the match between donor and acceptor energies. Build the antenna too sparse and the handoffs become slow and lossy. Build it too dense and new problems appear: excitations can become trapped, or multiple excitations can collide, leading to wasteful pathways. Architecture is not about maximizing one parameter; it is about balancing several that fight one another.

A concrete example helps. The major light-harvesting complex in green plants, often called LHCII, sits in the thylakoid membrane of chloroplasts and is among the most abundant membrane proteins on Earth. In the early 2000s, structural biologists were able to solve high-resolution structures of this complex, revealing a compact protein scaffold holding a small crowd of pigments: a mixture of chlorophyll *a* and chlorophyll *b*, plus carotenoids. The remarkable feature is not simply that pigments are present, but how tightly and asymmetrically they are arranged. Some chlorophylls sit close enough that their electronic states influence each other; others are positioned to act as stepping stones between regions. Carotenoids are threaded through the complex in ways that allow them to accept energy from chlorophylls under dangerous conditions and dissipate it harmlessly as heat. The same architecture that harvests light can, under excess light, shift into a protective mode. The antenna is not a passive bowl catching photons; it is a responsive device whose geometry makes photoprotection physically possible.

Long before these details were visible, the existence of such antenna systems was inferred from a more old-fashioned observation: photosynthetic organisms can be very efficient at low light. In the 1930s and 1940s, as quantitative biophysics began to infiltrate biology, researchers noticed that the number of chlorophyll molecules per reaction centre was large. That ratio implied an antenna: many chlorophylls feeding one conversion site. The idea matured as spectroscopy improved. Pigments were found to have subtly different absorption peaks, and energy transfer between them could be tracked by the way fluorescence shifted over time. Even without atomic structures, the inference was clear: the photosynthetic membrane is organized, not random.

Bacteria push the architecture into even more distinct forms. Purple photosynthetic bacteria build light-harvesting complexes that resemble rings. One famous example, known as LH2, forms a circular arrangement of repeating protein subunits. The pigments embedded in this ring include bacteriochlorophylls arranged so closely that they behave less like independent light absorbers and more like a collective. In schematic terms, the ring is an optical antenna wrapped around an empty centre, and energy is passed inward to another complex (LH1) that surrounds a reaction centre. The geometry is elegant: circular symmetry gives robust coupling; the dense packing allows rapid energy sharing; the layered arrangement creates a natural pathway to the reaction centre. This is architecture as topology: the path is not a literal wire but a designed network.

The difference between plant antennas and bacterial rings is not merely aesthetic. It reflects different lifestyles and constraints. Plants must survive violent swings in light intensity: dawn, noon, shade, cloud. They must endure oxygen-rich environments in which excited pigments can create destructive singlet oxygen. They therefore require flexible protection mechanisms, and their antennas are designed to switch between harvesting and quenching. Many bacteria live in more stable light environments, sometimes low light, sometimes filtered through water or other organisms. Their antennas can become more specialized for efficient capture and rapid transfer, and the geometry can become more regular.

At this point the "distance problem" comes into focus. Suppose a photon is absorbed by a chlorophyll in an outer antenna region. The excitation energy must reach a reaction centre in a few tens to hundreds of picoseconds, often faster, depending on the organism and conditions. That sounds leisurely until you remember that on the same timescale excited pigments can fluoresce, or lose energy to vibrations, or undergo harmful side reactions. The window is narrow. And the route is not empty space; it is a crowded membrane where proteins flex, water molecules tug, and thermal motion never stops.

A useful way to picture the challenge is to imagine building a relay race in a stadium during an earthquake. The runners are the pigments; the baton is the excitation energy. Each handoff depends on the runners being close enough and aligned enough to make contact, while the ground is shaking and the runners themselves are wobbling. The stadium, moreover, is partially filled with oxygen—a chemical spectator that, when struck by an excited runner, can turn violent. Under those conditions, efficiency is not a miracle of luck; it must be designed.

Design in biology is not conscious, of course, but it is real in the only sense that matters: the structures that persist are those that solve the physical problem. Proteins "tune" pigments by local electric fields, by small distortions of pigment shape, and by controlling which side groups are exposed to polar environments. The tuning is subtle but cumulative. If the reaction centre requires energy in a certain range, nearby pigments are adjusted to be good donors to it. If outer pigments are meant to absorb blue or green light, their environment is arranged accordingly. The result is a spectral organization across space: colour-coded plumbing.

A schematic of this energetic plumbing helps more than a photograph of atoms, because the key idea is a flow, not a literal funnel-shaped object.

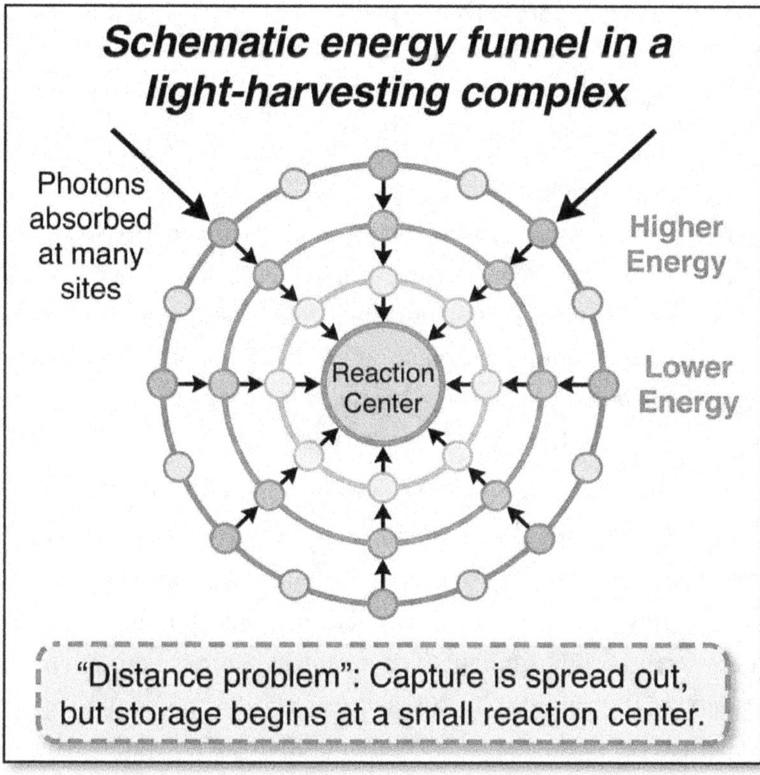

The diagram hides the most counterintuitive fact about the architecture: the energy does not travel as a tiny glowing marble moving from pigment to pigment in a straight line. Even in a strictly classical picture, energy transfer is mediated by electromagnetic interactions between molecules, not by physical migration of a particle. In the real molecular setting, the interactions are so sensitive to arrangement that the protein scaffold becomes as important as the pigments themselves. The scaffold is the reason the antenna is not just a dye solution. If you dissolve chlorophyll in a solvent, it absorbs light; it also wastes it quickly and often degrades. Put chlorophyll in a protein pocket and suddenly it becomes part of an information-processing network: energies are selected, pathways are biased, and dangerous side reactions are suppressed.

This is why the phrase "light-harvesting complex" can mislead. Harvesting is not simply collecting; it is collecting *and delivering*. Delivery, in turn, is not simply proximity; it is controlled coupling. A pigment that is too close to a neighbour might share energy so strongly that an excitation becomes spread in a way that no longer matches the reaction centre; a pigment too far might become an isolated dead-end. The architecture must place pigments in an intermediate regime where transfer is fast but not chaotic. And it must do so while embedded in a soft, fluctuating membrane.

The practical consequences show up in things people actually see. The deep green of a healthy leaf is not just the colour of chlorophyll absorption; it is also a compromise. If leaves absorbed *all* visible wavelengths strongly, they would overheat and produce damaging chemistry in bright sunlight. The antenna system is built not merely to maximize capture, but to match capture to processing capacity. Under excess light, plants activate non-photochemical quenching: a set of mechanisms that redirect excitation energy into harmless heat. Those mechanisms are enabled by the architecture of LHCs and their accessory pigments. In other words, the antenna is designed not only for famine (dim light) but also for feast (bright light), and the same structural elements that enable efficient transfer also enable safe disposal.

A historical anecdote from outside biology makes the point in a different key. When Otto Warburg, in the early twentieth century, measured photosynthetic oxygen evolution with painstaking care, he helped turn photosynthesis into a quantitative science, something that could be interrogated with rates and efficiencies rather than described only in botanical language. Later, as spectroscopic tools improved, the antenna idea became unavoidable: the numbers demanded it. A single reaction centre could not plausibly account for the absorption cross-section implied by the observed rates; many pigments had to feed it. Decades after Warburg, when membrane-protein crystallography finally made such complexes visible in atomic detail, the structures did not merely confirm the antenna concept; they revealed how elaborate the solution was. Biology had not scattered pigments like confetti. It had arranged them like circuitry.

Calling it circuitry should not imply rigid wires. The photosynthetic membrane is more like a crowded marketplace than a clean laboratory bench. Proteins jostle. Lipids flow. Water penetrates in thin fingers. A light-harvesting complex must maintain a stable geometry despite all of that, and it does so using the same forces that stabilize all proteins: hydrophobic packing, hydrogen bonds, and the snug fit of pigments into tailored pockets. The pigments are not simply glued in place; they are held by many small interactions that, collectively, are hard to disrupt. That is another aspect of the architecture: robustness. A leaf has to work after being buffeted by wind, heated, cooled, and partially dehydrated. The Sun is consistent; the environment is not.

There is also a more subtle robustness built into the architecture: redundancy of pathways. If energy transfer had to follow a single narrow route, a small perturbation—a change in pigment energy due to temperature, a slight shift in protein conformation—could block it. Many light-harvesting systems instead resemble networks with multiple viable routes leading inward. This does not require a conscious plan; it is an outcome favored by selection because it works under variable conditions. It also softens the distance problem. If energy can move along many partial routes, it is less likely to become stranded.

Seen from this angle, photosynthesis begins to look less like a single reaction and more like a managed supply chain. Photons are intermittent and unpredictable; excitations are perishable; the conversion centre has limited throughput; and the whole operation must avoid catastrophic side reactions. The light-harvesting complex is the warehouse and logistics hub, designed to keep energy moving, keep it safe, and deliver it in the right form to the place where chemistry can pin it down.

By the time the excitation reaches the reaction centre, it has already been filtered and shaped by this architecture. It is no longer merely "energy from the Sun"; it is energy that has survived a gauntlet of possible losses, guided by a scaffold built from ordinary biological matter—amino acids, pigments, and lipids—yet tuned with extraordinary precision. A sun trap, in the end, is not a net. It is a built landscape that makes the right thing easy and the wrong thing hard, so that a flicker of light can become, reliably and again and again, the starting pulse of life's chemistry.

3.2 The Exciton Relay

A photon vanishes the moment it is absorbed. Nothing luminous travels onward, no tiny pellet of sunlight rolling through the cell. What remains is a particular kind of disturbance inside matter: an electron has been lifted into an excited state, leaving behind an absence that behaves, in many respects, like a positive charge. In everyday language we might still say "the energy moves," but in the molecular world it matters *how* it moves, because different kinds of motion decay in different ways.

The classical picture that suggests itself is the one we learn early: a molecule gets excited, then passes that excitement to a neighbour, which passes it to the next, and so on. It resembles a bucket brigade. Each bucket handoff is a discrete event, and the bucket is always held by one person at a time. If that were the whole story, then the only question would be whether the handoffs are fast enough and frequent enough to outrun loss.

Nature does use something that looks like this in many contexts, and the mathematics behind it can be very successful. Yet in the dense pigment networks of photosynthesis, the bucket-brigade metaphor begins to mislead. Pigment molecules can be close enough, and their electronic transitions similar enough, that an excitation is not confined to one molecule in the way a bucket is confined to one pair of hands. Instead, the excited state can be shared among several pigments at once. This shared excitation is called an *exciton*.

An exciton is not a new particle in the sense that an electron is a particle. It is better thought of as a coordinated pattern of electronic excitation spread over a cluster of molecules. If you pluck one string on a guitar, the vibration is mainly on that string. But if the strings are coupled and the wood resonates strongly, a single pluck can set up a vibration pattern that belongs to the whole instrument. The energy is still localized to the guitar, but it is not localized to a single string. In a light-harvesting complex, pigments are the strings and the protein scaffold is the resonant body that holds their positions and shapes their coupling.

This idea can feel like a semantic trick until one sees what changes when the excitation is shared. In a bucket brigade, the energy has an address: molecule A, then molecule B, then molecule C. In an exciton picture, the energy has a *distribution*. One excitonic state might be mostly on pigment A but partially on B and C; another state might be spread more evenly; another might be concentrated in a different region of the complex. The relay does not only involve movement from one pigment to another. It also involves transitions between these shared states, and those transitions can be guided by the environment in ways that are hard to see if one insists on the bucket.

A useful starting point is the force that couples pigments to each other. When a chlorophyll absorbs light, its electrons rearrange, creating a changing electric dipole moment. A neighbouring pigment "feels" that changing electric field. If the pigments are close and well oriented, the field can drive the neighbour into its excited state without a photon being emitted. This is resonance energy transfer, a process long studied in chemistry. In the simplest regime, where pigments interact weakly, you can treat transfer as a hop: excitation on one molecule, then on the next. In the stronger-coupling regime, where the interaction energy is comparable to differences between pigment energies and to environmental fluctuations, the language of hops becomes strained. The system prefers to organize itself into collective excited states: excitons.

The word itself carries a bit of twentieth-century physics history. It was coined in solid-state physics, where people were trying to describe what happens when light excites an electron in a crystal and leaves behind a positive "hole." The electron and hole can attract each other and move together as a neutral composite: an exciton. In molecular aggregates like photosynthetic antennas, the details differ—the electron and hole are not roaming freely through a crystal lattice—but the central idea survives: excitation is a delocalized entity with its own dynamics.

This is where the quantum part of the story becomes unavoidable, but it need not become mystical. "Quantum" here simply means that the excitation is described by a wavefunction, and that wavefunctions can spread across multiple sites. The surprising feature is not that quantum mechanics applies—it always does at the scale of molecules—but that the sharing can be extensive enough to matter for function in a warm, wet, noisy cell.

Consider the purple bacterial ring complexes mentioned earlier. In these rings, bacteriochlorophyll pigments are arranged with near-regular spacing and strong coupling. The collective excited states of such a ring can be quite delocalized: an exciton can wrap around multiple pigments at once. In that setting, the antenna does not merely present many independent absorbers. It presents a cooperative object with absorption properties that can differ from any individual pigment. Some exciton states couple strongly to light (they are “bright”), while others couple weakly (they are “dark”). Bright states are good for capturing photons; dark states can be useful as temporary hiding places where the energy is less likely to be re-emitted as fluorescence. Architecture makes these states available.

Plants and algae offer a less symmetric but equally instructive example. Their antenna complexes pack chlorophyll *a* and *b* at various distances and orientations. Some groups of pigments are close enough to form strongly coupled clusters. Within such a cluster, it often makes more sense to talk about exciton states than about excitation sitting on one chlorophyll. Spectroscopists see this in the way absorption bands split and shift compared with isolated pigments. When pigments talk to each other strongly, they no longer sing exactly the same note.

A person at a dinner party might reasonably ask: why should sharing the excitation help? If energy must end up at a reaction centre anyway, why bother with this extra layer of description? One answer is simply honesty: the exciton picture is what the physics demands when couplings are strong. But there is also a functional intuition.

In a bucket brigade, if one handoff is slow or blocked, the relay stalls. In a delocalized excitation, the system can, in effect, “test” more than one route at once. That sentence can easily be overinterpreted, as though the excitation were a little creature making choices. It is not. What it means is that when an exciton spans multiple pigments, the probability of finding the excitation on any one pigment is distributed across the span. If one pigment is well connected to the next stage in the energy funnel, there is automatically some amplitude already sitting near that exit. Transfer can then occur without waiting for a long sequence of discrete hops to bring the excitation to precisely the right spot.

There is also a subtler point: delocalization can change how the environment interacts with the excitation. A single localized excitation couples strongly to local vibrations and local fluctuations; it can lose coherence quickly and relax into heat. An exciton spread over several pigments can average over some of that noise, at least briefly, in the way that a choir can hold pitch better than a single singer in a windy street. This does not mean the system becomes immune to noise; it means noise shapes the dynamics in a richer way than mere disruption.

The word *coherence* is often used here, and it deserves care. In quantum mechanics, coherence refers to the maintenance of well-defined phase relationships between parts of a wavefunction. If an exciton is in a coherent superposition of being on pigment A and pigment B, there is a relative phase between those possibilities. That phase can produce interference effects, which can accelerate or redirect energy flow, the way water waves can add or cancel depending on how their crests align. In the messy environment of a cell, phases tend to wash out quickly. The process by which they wash out is decoherence, driven by interactions with the surrounding vibrations and fluctuating electric fields of the protein and solvent.

Here a historical episode enters, because excitons in photosynthesis moved from being a respectable theoretical convenience to being a public symbol of "quantum life" largely through a striking set of experiments. In 2007, Gregory Engel and colleagues used two-dimensional electronic spectroscopy to probe a photosynthetic complex from green sulfur bacteria called the Fenna—Matthews—Olson (FMO) complex. Their measurements revealed oscillatory signals—quantum beats—that suggested coherent superpositions among excitonic states. The result was thrilling not because it proved quantum mechanics (nobody doubted that), but because it suggested that coherence might persist long enough to influence biological function, even in a warm, noisy setting.

The technique itself has a certain drama. Two-dimensional electronic spectroscopy uses sequences of ultrafast laser pulses to excite and interrogate a sample, producing spectra that correlate excitation and emission frequencies over time. It is a bit like striking a bell with carefully timed taps and listening not just to the notes but to how the overtones exchange energy. When coherence is present between exciton states, it can show up as oscillations in the signal as a function of the waiting time between pulses. These oscillations can be interpreted as the system's quantum phases marching in step for a while before being blurred by the environment.

The headlines that followed were, perhaps inevitably, exuberant. "Quantum coherence in photosynthesis" became a phrase that migrated into popular accounts and grant proposals. Yet the scientific story quickly became more nuanced. Subsequent work, including careful analyses of what the spectroscopy signals actually imply, argued that at physiological temperatures electronic coherence may decay very quickly, on the order of tens of femtoseconds. Some long-lived oscillations might arise not from long-lived electronic coherence but from vibrational coherences or from mixed vibronic effects—a coupling between electronic states and specific vibrations—or from subtleties in how the signals are generated and interpreted. Hong-Guang Duan and Valentyn Prokhorenko, among others, pushed this more orthodox view, emphasizing that efficient energy transfer does not require long-lived electronic coherence.

This might sound like a deflation, but it is better seen as a clarification of what is really interesting. The core question is not whether coherence lasts for a picosecond or a tenth of a picosecond in a particular complex under a particular temperature. The deeper question is how energy transfer in these systems is optimized in the presence of noise, and whether quantum mechanics offers design principles that would be invisible in a purely classical model.

One such principle has become famous because it is counterintuitive: a *moderate* amount of noise can help. In 2008, Martin Plenio and Susana Huelga developed theoretical models showing that dephasing—the very process that destroys coherence—can, in certain network structures, *increase* the efficiency of transport to a sink (a reaction centre analogue). Too little noise can allow the excitation to become trapped by quantum interference or localization effects; too much noise turns transport into slow, classical hopping. In between lies a sweet spot where fluctuations shake the system just enough to keep it exploring pathways that lead inward.

This is sometimes called environment-assisted quantum transport. The phrase is a mouthful, but the idea is simple enough to hold in the mind: the environment is not merely an enemy. The protein's vibrations and fluctuations can act like a helpful jostling crowd, preventing the excitation from getting stuck in a cul-de-sac. In a city analogy, coherence is like a perfectly smooth, frictionless traffic system where cars can, bizarrely, end up circulating in loops because of the symmetry of the roads. A bit of friction and randomness breaks the symmetry and lets traffic drain toward the highway exit.

Seen this way, the exciton relay is not a binary choice between "quantum wave" and "classical hop." It is a spectrum of behaviours determined by coupling strengths, energetic disorder, and environmental noise. The pigment network provides the couplings; the protein scaffold and surrounding solvent provide the noise; the overall architecture sets the energetic slope toward the reaction centre. The exciton is the language that lets one talk about this interplay without forcing everything into the metaphor of a baton.

There is a further reason excitons matter, and it returns us to the practical concern that started the story: loss. If an excitation sits localized on a pigment that fluoresces readily, it can be lost by emission. If it sits on a pigment that is prone to harmful chemistry, it can trigger damage. In a network with exciton states, the system can rapidly relax from bright, highly absorbing states into darker states that are still able to transfer energy but are less likely to radiate it away. That is not magic; it is simply what happens in coupled systems with multiple eigenstates. But it is functionally significant.

At the same time, there is a limit to how far one should romanticize the exciton. Delocalization over a handful of pigments is common in tightly packed complexes; delocalization over dozens in a warm, disordered protein is harder. Excitons are not necessarily huge, pristine waves. They can be modestly spread, continually buffeted, partially localized, and frequently reshaped by their environment. The temptation in popular accounts is to imagine a single coherent wave sweeping flawlessly to the reaction centre like a perfectly aimed whisper. The reality is more like a chorus in a crowded room: the harmony exists, but it is constantly negotiated with the acoustics.

Even so, the exciton concept changes the way photosynthesis looks. The antenna is not merely a collection of independent pigments; it is a coupled quantum object whose excited states are collective. The relay is not merely a chain of handoffs; it is a sequence of transformations among shared excitations, guided by an energy landscape and modulated by noise. The light-harvesting complex is therefore not only a solar collector but also a signal-processing device: it takes random photon hits and converts them into a directed flow of excitation that can be delivered, with remarkable reliability, to the molecular machinery that turns fleeting light into lasting chemical separation.

That reliability is what makes fields green and oceans productive. It is what allows a plant to thrive under a patchwork of sun and shade, and a bacterium to make a living from the faintest filtered glow. Somewhere in the crowded interior of a protein scaffold, an excitation becomes less like a possession carried by one molecule and more like a shared disturbance that finds its way by being, briefly and usefully, in more than one place at once.

3.3 The Wire in the Deep

A great deal of photosynthesis happens where you would least expect it: not in bright meadows or shallow water, but in places where light is a rumor. Green sulfur bacteria can live in environments so dim that a plant would starve. Some thrive in deep, murky water where sunlight has been sifted by distance and by the appetites of everything above them. They are not simply enduring low light; they have built a way of drinking it, capturing and using photons so scarce that each one counts.

Under such conditions the familiar problems of light harvesting become brutal. In bright sunlight, a plant can afford occasional inefficiency because another photon will arrive a moment later. In near-darkness, missed opportunities are existential. The antenna must be enormous, because the chance of intercepting a photon is tiny. But a huge antenna brings a new kind of vulnerability: it spreads the initial capture far from the place where energy can be stabilized. The energy must travel farther, through more molecular clutter, with more chances to leak away. A long-distance transfer problem appears, and biology answers it with a component that behaves, in spirit, like a wire.

The protagonist is the Fenna–Matthews–Olson complex, usually abbreviated to FMO. It is found in green sulfur bacteria, sitting between a gigantic light-harvesting structure called the chlorosome and the reaction center where charge separation begins. The chlorosome is a marvel in its own right: a packed aggregate of bacteriochlorophylls that can absorb faint light efficiently. But that very aggregation creates a challenge. The chlorosome is like a vast satellite dish bolted to the outside of the photosynthetic membrane, while the reaction center is embedded inside the membrane and demands delivery at a particular site. FMO is the bridge, a pigment–protein complex that mediates the transfer from the bulky antenna to the reaction center with high reliability.

Calling it a bridge risks underselling the precision of the object. FMO is not a general connector. It is a structured network of pigments held in a protein scaffold, arranged so that excitation energy tends to flow in a preferred direction. It is also compact: a little molecular "module" that can be repeated and packed efficiently. The sense in which it is a wire is not that an electron travels down it (that happens later, in the reaction center), but that excitation energy is guided through it, from one end to the other, before it has time to disappear.

The key ingredients of this guidance are few but powerful. First, the pigments inside FMO are bacteriochlorophylls positioned close enough to interact strongly, forming shared excitations (excitons) rather than isolated hops. Second, the protein environment tunes the energies of those pigments so that some are slightly higher in energy and others slightly lower. The downhill slope is gentle but consequential. Third, the complex sits in a particular orientation with respect to its partners: it "faces" the chlorosome on one side and the reaction center on the other, so that the energetic landscape is also a spatial one.

A schematic can say this more clearly than words. FMO is often described as containing seven bacteriochlorophyll pigments in its best-known form, arranged inside a protein scaffold like beads embedded in a translucent jelly. The scaffold does not merely keep them from drifting away; it fixes their separations and angles, which determine how strongly they couple and how the excitation can spread.

FMO complex (schematic): seven bacteriochlorophylls in a protein scaffold

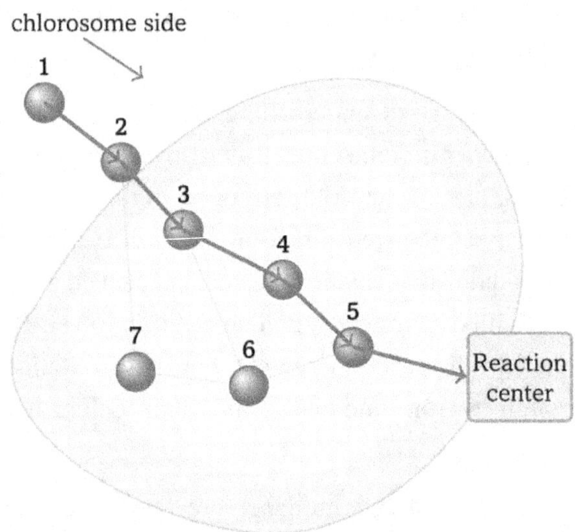

Protein scaffold tunes pigment energies and couplings; excitation tends to drift toward the reaction center.

In real FMO, the geometry is more intricate than this sketch, but the logic is captured: a network of pigments, embedded in protein, with a preferred direction of delivery. The payoff of this design becomes clearest when one remembers what the reaction center is waiting to do. Harvesting light is not, by itself, useful.

The crucial act is to convert the fleeting excitation into a more durable form: a separated positive and negative charge that can be used to drive chemistry. Reaction centers accomplish this through a choreographed sequence of electron transfers along a chain of cofactors—special pairs of chlorophyll-like molecules, accessory pigments, pheophytins, quinones—all held at specific distances and orientations. The sequence is fast, directional, and energetically downhill. The spatial separations are typically too large for a single leap of electron tunneling to be reliable; the chain exists because long-range transfer in proteins is brutally distance-dependent. Biology solves the distance problem, again, with stepping stones.

FMO sits just upstream of that. It delivers excitation to the reaction center in a form the reaction center can accept, and on a timescale that keeps it from being wasted. In near-darkness, this delivery must be exceptionally dependable, and the bacterial lifestyle has pushed the architecture toward that dependability.

The story acquired a modern twist because FMO became a testing ground for an unusually vivid question: could a biological system exploit quantum coherence to enhance performance? That question was not asked in a vacuum. Quantum mechanics is already in every chemical bond; the question is whether *coherent* quantum dynamics, the wave-like phase relationships that can produce interference, can persist long enough and over enough pigments to matter for the flow of energy.

FMO attracted attention partly because it is relatively small and structurally well-characterized, making it tractable for theory, and partly because green sulfur bacteria are such extreme specialists that one expects their machinery to be pushed close to physical limits. In 2007, Engel and colleagues reported oscillatory signatures in two-dimensional electronic spectroscopy of FMO at cryogenic temperatures (around 77 K). Those oscillations were interpreted as quantum beating among excitonic states—a kind of coherent sloshing of excitation across the network. The image was irresistible: in a biological "wire," energy might travel not as a random walk but as a coordinated wave exploring multiple pathways.

As with most irresistible images, the reality became richer and more contentious with time. Later work emphasized that at physiological temperatures the electronic coherences may decay very rapidly, often within tens of femtoseconds, and that some long-lived oscillations can arise from vibrations or mixed vibronic effects rather than long-lived electronic superpositions. The scientific value of this debate is not in deciding whether nature is "quantum" (it is), but in pinpointing what aspects of quantum dynamics survive in a warm protein and how they shape function. Even if long-lived electronic coherence is not the main driver at ambient temperature, excitonic coupling is still real, and it still affects rates and pathways. The system can be quantum without being a fragile quantum computer.

One idea that emerged from theoretical work is that FMO, like other light-harvesting networks, may operate in a regime where neither perfect coherence nor complete incoherence is ideal. In that intermediate regime, environmental noise—the jitter of the protein, the jostle of surrounding water, the vibrations of pigments—does not simply destroy useful dynamics. It can also prevent the excitation from becoming stuck in patterns of interference or in localized traps. A modest amount of dephasing can help the system keep sampling routes that lead to the reaction center. This is a subtle kind of robustness: the same restless thermal environment that threatens to erase delicate quantum phases can also keep the energy moving.

To appreciate why a small "wire" like FMO might benefit from such a balance, imagine the excitation arriving from the chlorosome. It is not delivered with a label saying which pigment to occupy. It arrives as a probabilistic event, and the initial excitation might land in a superposition of excitonic states, or it might rapidly relax into a subset of them. From there, the network must guide it to the sink. If the network were perfectly symmetric and perfectly coherent, interference could, in principle, keep some amplitude circulating away from the exit. Real proteins are not perfectly symmetric. Energetic disorder and environmental fluctuations break symmetries naturally. The fascination is that the disorder is not merely a flaw; it is an ingredient that can make the transport reliable under varied conditions.

The "wire" metaphor becomes especially illuminating when one contrasts FMO with the antenna. In the chlorosome, the challenge is capture: maximize absorption in dim light. Dense packing and large aggregates accomplish that, but at the cost of complexity. In the reaction center, the challenge is conversion: create charge separation rapidly and directionally. In between, FMO's task is delivery: take excitation from a messy, large antenna domain and hand it to a precisely tuned conversion machine. A wire in a piece of electronics does not generate energy; it reduces losses between generator and load. FMO does something analogous at the molecular level.

The stakes of that reduction are not academic. In low-light habitats, organisms are competing not for territory but for photons. A slight improvement in the probability that an absorbed photon leads to charge separation can be the difference between slow growth and no growth. There is a quiet drama in imagining a microbial mat in deep water: layers of organisms stacked like living filters, each consuming a portion of the remaining light. Those at the bottom inhabit a world of tinted darkness, living on the scraps of the spectrum that penetrate. In that world, FMO is not a curiosity; it is a livelihood.

The name Fenna–Matthews–Olson hints at another human story: structural biology, and the patience required to see what nature has built. Before lasers and ultrafast spectroscopy could interrogate energy flow, crystallography and biochemical purification made it possible to isolate and describe the physical components. The identification and characterization of FMO as a discrete complex gave later researchers a manageable object: small enough to model, stable enough to measure, and important enough to matter. When Engel's group brought ultrafast tools to it decades later, the complex was ready to become a meeting place for disciplines that do not always share a language: microbiology, physical chemistry, quantum theory, and spectroscopy.

What survives beyond any particular experimental controversy is the basic design lesson. In photosynthesis, energy is precious not because it is rare in principle, but because it is perishable in practice. The architecture that catches light must be coupled to an architecture that stabilizes it, and the gap between the two is not bridged by wishful thinking. It is bridged by proteins that hold pigments in just the right places, tuning their energies, shaping their couplings, and allowing a brief, useful form of shared excitation to carry the day.

In the dim habitats of green sulfur bacteria, the wire is short, but the distance it spans is measured against the lifespan of an excitation. That is the relevant metric. FMO makes that distance effectively smaller, not by shrinking space, but by arranging matter so that the excitation finds the reaction center quickly, before the surrounding warmth can take back what the photon briefly offered.

Chapter 4

Surfing the Noise

In the chaotic warmth of a living cell, delicate quantum states should vanish almost instantly, victim to the relentless thermal storm. Yet, evidence suggests that photosynthesis doesn't merely survive this environment but thrives on it, utilizing molecular vibrations to shake energy loose from traps and guide it toward its destination. This chapter reveals how life performs a thermodynamic jujitsu move, turning the enemy of quantum mechanics-noise-into an ally.

4.1 The Rhythm of the Exciton

Sunlight arrives in discreet packets of energy called photons, each one a tiny down payment that must be spent quickly. In a leaf, a bacterium, or an algae cell, that energy cannot be wired straight into chemistry the way a wall socket powers a kettle. It has to be handed from molecule to molecule until it reaches a reaction centre, the specialised site where charge separation begins and the long business of metabolism can start. If the handoff is too slow, the energy leaks away as heat or harmless glow; if it is too clumsy, it triggers damaging side reactions. The everyday miracle is that photosynthetic organisms manage this handoff at breathtaking speed and with striking efficiency, in a world that is warm, wet, and busy with molecular jostling.

The first surprise, which chemistry students meet early, is that the relevant unit of energy transfer is not always a particle-like "excited molecule" hopping between well-separated pigments. Instead, when pigments sit close enough and their electronic states interact, the excitation can be shared: it becomes an *exciton.* An exciton is an electronic excitation that is delocalised over more than one pigment, a collective state of the whole pigment cluster. The exciton is not a new particle floating around; it is a bookkeeping device for the fact that the excitation is, in a quantum sense, smeared across a network.

That smearing matters because the network is not a simple line. Light-harvesting complexes are crowded molecular scaffolds: pigments embedded in proteins, arranged with the particularity of architecture rather than the randomness of a soup. Their geometry creates a landscape of couplings and energy gaps. To reach the reaction centre quickly, the exciton must navigate that landscape. The question that sharpened in the early 2000s was deceptively simple: does the exciton move through that landscape in a manner that retains quantum phase relationships long enough to make a difference? Put less formally: does it behave like a wave long enough to choose good routes by interference, or does it collapse almost instantly into a classical hop?

By the time the story reached the wider scientific imagination, it had acquired an alluring emblem: *quantum beats.* These are oscillations in an optical signal that look, at first glance, like the ticking of a microscopic metronome. The claim was not merely that photosynthetic complexes are quantum (everything is), but that a measurable, coherent quantum rhythm persists under conditions far closer to biology than many physicists would have wagered.

A rhythm, of course, is only as convincing as the instrument that hears it. The key advance was a class of techniques that can be described, without too much violence to the details, as ultrafast cinematography for electronic motion. Ordinary spectroscopy shines light and measures what is absorbed or emitted, giving a spectrum: intensity versus wavelength. That is like taking a still photograph. But the exciton's most interesting life unfolds on femtosecond timescales. A femtosecond (10^{-15} s) is to a second what a second is to about 32 million years. If energy transfer between pigments can happen in a few hundred femtoseconds, then the relevant processes are over before most molecular vibrations have even completed a cycle.

Two-dimensional electronic spectroscopy (2DES) was developed to resolve this. The name sounds forbidding, but the underlying idea can be made approachable. Instead of using a single pulse of light, 2DES uses a carefully timed sequence of ultrashort laser pulses. You can think of the first pulse as "writing" information into the system (creating a coherent superposition of electronic states), the next as allowing the system to evolve for a controlled delay, and a later pulse as "reading" out what has happened by stimulating an emitted signal. By scanning delays and analysing the emitted light, one reconstructs a two-dimensional map: how excitation at one frequency is correlated with emission at another, and how these correlations evolve over time.

Why two dimensions? Because the energy landscape in a pigment-protein complex is crowded. Many transitions overlap. In a one-dimensional spectrum they blur together; in two dimensions they can separate into peaks whose positions and couplings tell a richer story. Crucially, the technique can reveal not just populations (how much excitation is in a given state) but also *coherences*—phase relationships between states. Coherence is the mathematical statement that the system occupies a superposition like $\alpha|A\rangle + \beta e^{i\phi}|B\rangle$, with a definite relative phase ϕ. If that phase drifts randomly due to environmental noise, coherence decays (decoherence). If it persists, it can create interference effects: the hallmark of wave-like dynamics.

The historical anecdote that made the rhythm famous centres on a particular photosynthetic complex from green sulfur bacteria: the Fenna–Matthews–Olson (FMO) complex. FMO is not the primary light-harvesting antenna; it is more like a molecular wire that connects a large antenna (the chlorosome) to the reaction centre. It contains a small number of bacteriochlorophyll pigments arranged within a protein scaffold. This modest size, compared with sprawling plant antennas, made FMO an appealing testbed: complicated enough to be biologically meaningful, simple enough to be modelled.

In 2007, Gregory Engel's group reported 2DES measurements on FMO at cryogenic temperature (around 77 K, the temperature of liquid nitrogen). The headline result was the observation of oscillatory features in the 2D signal as a function of the waiting time between pulses. These oscillations were interpreted as quantum beats between excitonic states: the system was coherently superposed across exciton energy levels, and the phase difference between them evolved at a frequency set by their energy splitting. In ordinary language, the exciton seemed to ring like a struck bell, with a beat pattern that could be timed and tracked.

To see why beating implies a superposition, consider an analogy closer to sound. If two musical notes of nearby frequency are played together, the resulting sound swells and fades at the difference frequency: the familiar acoustic beat. In quantum mechanics, a superposition of two energy eigenstates evolves in time with two different phase rates, so observables can oscillate at their energy difference. The oscillation is not an optional aesthetic feature; it is a signature that the state retains phase information connecting the two energies. If the system immediately lost that phase information to its environment, the beat would wash out.

What made the report electrifying was not that beats exist in quantum systems—they are routine in carefully isolated atoms and superconducting circuits—but that they appeared in a pigment-protein complex, a soft biological material. Even more tantalising, variations of the experiment suggested that oscillations could persist on timescales that seemed long compared with naive expectations. This fed a bold narrative: perhaps photosynthesis uses coherence to explore multiple pathways simultaneously, sampling the network as a wave would, and thus finding an efficient route to the reaction centre.

That narrative was always in danger of being oversimplified, and the later debate about what, precisely, is oscillating is important. But it is worth lingering on why the original observation mattered even before the controversy. The achievement was not merely interpretive; it was experimental. Ultrafast spectroscopy is as much engineering as physics: generating pulses only a few tens of femtoseconds long, controlling their phases and delays, stabilising interferometers against thermal drift, and extracting weak nonlinear signals from noise. In a laboratory, the apparatus looks less like a microscope and more like a carefully choreographed obstacle course for light: mirrors, beam splitters, delay stages, and crystals, all arranged so that pulses meet a sample with exquisite timing.

The sample itself is, from a physicist's perspective, almost rude. It is not a single molecule trapped in vacuum, but an ensemble of complexes in solution or in a glassy matrix, each one slightly different due to thermal disorder and the quirks of protein conformations. That disorder broadens spectral lines and tends to erase delicate phase information. The reason 2DES is powerful is that it can separate different contributions by their distinct signatures in the 2D map and by their dependence on pulse polarisation and timing. Even so, interpreting the maps is closer to forensic analysis than simple reading.

At the heart of the beat story is the distinction between *population* and *coherence*. Population is like money sitting in a particular bank account: the excitation is on a certain pigment or in a certain excitonic state. Coherence is like having money in two accounts with a fixed exchange rate and a contract connecting them; it is not merely "some in each," but "both at once in a phase-related way." 2DES can be sensitive to coherence because the emitted signal depends on how the system's polarisation—a collective oscillation of charge induced by light—evolves between pulses. When two excitonic states remain phase-related, that polarisation can show oscillations at the frequency corresponding to their energy gap.

A concrete picture helps. Imagine two pigments coupled so that their excitations mix. The coupled system has two excitonic states: one slightly higher in energy, one lower. If a pulse creates a superposition of these excitonic states, then the electronic charge distribution in the complex oscillates between patterns associated with each state. This oscillation is not necessarily a literal electron sloshing back and forth like water in a tube; it is a phase evolution in the quantum state that shows up in how the system interacts with light. The 2DES signal can carry that oscillation as an amplitude that rises and falls, producing the observed beats.

Even at this level, an important subtlety enters. When an oscillation appears in a spectroscopic signal, it does not automatically mean that the excitation itself is physically taking a wave-like route through the network in a way that improves transport. Spectroscopy measures how a system responds to controlled optical perturbations; biology runs on sunlight, which is incoherent on femtosecond scales. The pulses used in 2DES can prepare states that are more coherent than sunlight would. This does not make the measurement irrelevant, but it means the interpretation must be cautious: long-lived beats in a nonlinear optical signal are evidence that the system *can* support certain coherences under certain preparations, not necessarily that it *uses* them in the same way in nature.

The strongest, most defensible significance of the 2007-style observations was narrower and more robust: the pigment-protein complex exhibits coherent dynamics that survive nontrivially long in a condensed-phase environment, at least under some conditions. That is already remarkable. It forced theorists to abandon the simplest extremes. The classic picture of excitation transfer in many contexts is Förster resonance energy transfer (FRET), which treats transfer as incoherent hopping mediated by dipole–dipole coupling, valid when electronic couplings are weak compared with environmental reorganisation. At the opposite extreme is fully coherent wave-like motion governed by a static Hamiltonian. Real photosynthetic complexes live between these limits: couplings can be strong enough to delocalise excitations, but environmental fluctuations are strong enough to cause dephasing. The beats drew attention to that intermediate regime and to the need for models that treat system and environment together.

It also made the protein scaffold feel less like a passive holder of pigments and more like an active participant in the physics. Proteins are not rigid; they breathe. Their motion modulates pigment energies and couplings, producing a constantly shifting landscape. One might expect such motion to be purely destructive to coherence. Yet the beat observations suggested that, at least briefly, coherence can coexist with that motion. This set up a more nuanced question: rather than asking whether the environment kills coherence, ask how the environment shapes it, and whether certain patterns of fluctuation might protect or even assist specific dynamical features.

Placed near the Engel et al. experiment, a simplified sketch of what "beats fading over time" looks like can anchor the intuition. The essential shape is an oscillation whose envelope decays: the rhythm starts crisp and then becomes muffled as dephasing erases phase relationships. The decay is not merely experimental messiness; it is the physics of an open quantum system, where the exciton is coupled to a noisy molecular bath.

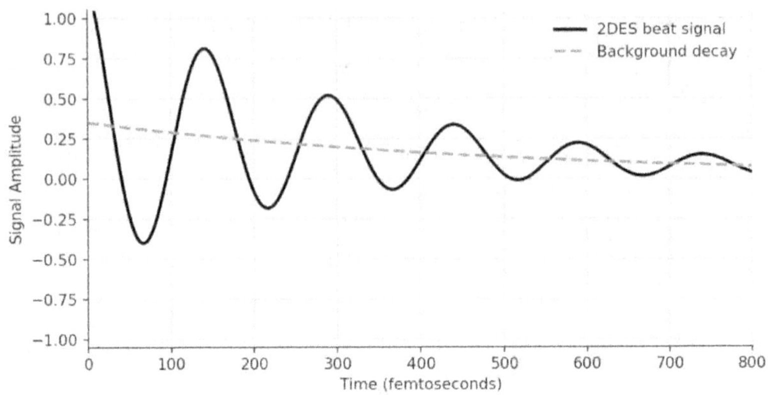

The beat frequency in such a plot corresponds to an energy gap: $\Delta E = \hbar\omega$. That connection is part of what makes the signal so tempting to interpret. If one can identify which excitonic states correspond to the observed frequencies, one can infer how the exciton is delocalised and how strongly different pigments are coupled. In favourable cases, the 2D maps show cross-peaks indicating coupling between transitions, and the oscillations can be associated with specific pairs of excitonic states. The result feels like eavesdropping on the complex's internal conversation.

Still, even the friendliest explanation must admit that a spectroscopic beat is a composite creature. The molecules involved are not simple two-level systems. Pigments have vibrational modes—ways the nuclei can move relative to each other—and these can couple to electronic excitation. Vibrations can themselves produce oscillatory signals. Some oscillations reflect purely vibrational coherence (nuclei moving in phase), some reflect electronic coherence, and some reflect *vibronic* mixtures where electronic and vibrational character intertwine. Disentangling these possibilities is difficult because they can share similar frequencies and decay times, and because the 2DES signal is generated by multiple quantum pathways in the light–matter interaction.

This is where the story becomes intellectually satisfying rather than merely seductive. The early reports did not settle a debate; they created one worth having. If the oscillations were electronic and long-lived, that would suggest that coherence might influence energy transfer in a way that is not captured by simple hopping models. If the oscillations were mostly vibrational, that would still be important: vibrations are part of the environment, and structured vibrational modes might couple to excitons in ways that alter transfer rates or pathways. Either way, the simplistic idea that warm, wet systems must immediately become classical looked less secure.

A second, quieter human element also matters. The field had long been divided by language. Photosynthesis researchers often spoke in the dialect of physical chemistry: rate constants, spectral densities, reorganisation energies. Quantum information theorists and condensed-matter physicists spoke of coherence, entanglement, and interference. The 2DES beat papers provided a shared object of fascination, a graph that both communities could point to. It brought theorists into closer conversation with experimentalists, and it forced both to be more precise about what counts as a meaningful quantum effect in a biological setting.

The exciton's rhythm, then, is not a claim that leaves are secretly quantum computers. It is a clue about timescales and couplings. It says that when light excites a tightly packed pigment network, the resulting excitation need not immediately choose a single pigment or a single pathway. For a brief window—brief, but not negligible on the scale of energy transfer—the excitation can exist in a superposition across excitonic states, and that superposition can leave an oscillatory fingerprint in ultrafast optical signals. The environment is present from the start, tugging at phases and reshaping energies, but it does not necessarily erase all quantum character instantaneously.

One can imagine the protein scaffold and its pigments as a small orchestra playing on a crowded street. The instruments begin in time, a clear rhythm emerges, and then the city noise swells: footsteps, engines, voices. The rhythm does not vanish at the first disturbance; it fades, sometimes quickly, sometimes more slowly, depending on how the instruments are arranged and how the street echoes. In photosynthetic complexes, the rhythm is measured not by ears but by laser pulses and emitted photons, and the fading is a record of how coherence yields to the push and pull of a living environment.

When the rhythm is audible, even briefly, it tells you something concrete: the excitation is not merely passing a parcel from hand to hand. It is, for a moment, a delocalised quantum state with an internal phase. That phase can beat, and the beat can be timed in femtoseconds, revealing that the boundary between "quantum" and "biological" is not a wall but a porous, experimentally accessible frontier.

4.2 The Goldilocks Shiver

A child on a playground quickly learns a rule that looks almost like wisdom: some pushing helps, too much pushing ruins the game. On a swing, a gentle nudge at the right moment builds a clean arc; a shove out of rhythm just makes the motion messy. Photosynthetic energy transfer lives under a similarly awkward rule. The exciton—that delocalised package of electronic excitement shared among pigments—would seem, at first glance, to crave perfect quiet. Quiet means coherence, and coherence means wave-like motion, and wave-like motion feels like the fastest way to get somewhere. Yet quiet can be a trap. A little vibration, a little thermal restlessness, can make the exciton *more* likely to reach its destination.

This is the core of environment-assisted quantum transport, usually abbreviated ENAQT. The acronym sounds like something assembled for grant proposals, but the idea is wonderfully concrete: the environment, often blamed for destroying delicate quantum behaviour, can also *rescue* quantum transport from its own worst habits.

The word *environment* here is not a vague philosophical backdrop. In a light-harvesting complex, it means the protein scaffold, surrounding water, and the internal molecular motions of the pigments themselves. These degrees of freedom jostle the electronic energies of the pigments, shifting them up and down in time. They also tug at the phase of a quantum superposition, causing *dephasing*: the gradual loss of the well-defined phase relationships that make interference possible.

Interference is the reason quantum transport can be so fast, but it is also the reason it can fail. The simplest demonstration is the double-slit experiment: a wave passing through two openings interferes with itself, creating bright and dark fringes. Those dark fringes are not mere absences; they are *cancellations*. A quantum amplitude can cancel another amplitude, preventing the particle from being found in certain regions. In a network of pigments, the exciton's probability amplitude can split and recombine along different routes. Sometimes that recombination helps: amplitudes add constructively toward the reaction centre. Sometimes it hurts: they cancel in precisely the wrong place, leaving the exciton marooned in a pattern that has little overlap with the exit.

The relevant kind of marooning has a technical name: trapping in an *invariant subspace*. The phrase sounds abstract, but the image is simple. Think of a network as a set of rooms connected by doors. If the rules of motion are perfectly coherent and perfectly symmetric, there can be combinations of rooms that talk only among themselves. You can wander forever inside that combination without ever finding the particular door that leads outside. This is not because the door is locked, but because the wave you are has a node there: the amplitude at the doorway is zero, by interference. The exit exists, but the coherent dynamics refuse to populate it.

This is the point at which “noise” becomes unexpectedly helpful. Dephasing noise does not simply smear everything into classical randomness. At moderate strength it does something more surgical: it breaks the strict phase relationships required for perfect destructive interference. It shakes the system out of the invariant subspace. Doors that were mathematically invisible become visible because the cancellation is no longer exact. The exciton gains access to pathways that coherence alone paradoxically denied.

The theoretical work that crystallised this idea, notably by Martin Plenio and Susana Huelga around 2008, framed transport through pigment networks in a language borrowed from quantum information: open quantum systems moving on graphs. In those models, you specify a Hamiltonian (the coherent couplings between sites or excitonic states), a sink (the reaction centre or charge-separation site where excitation is irreversibly captured), and an environment that causes dephasing. You then ask a bluntly biological question: what fraction of excitations reach the sink before they are lost to recombination or other decay channels?

When you turn the dephasing strength from zero upward, the models often show a curve with a peak. At zero dephasing, the system can be too coherent for its own good: interference traps part of the excitation away from the sink. At very strong dephasing, coherence is crushed so thoroughly that the system behaves like a classical random walk, or worse: it can be slowed by a quantum effect with a wonderfully ironic name, the *quantum Zeno effect*, where frequent "measurements" by the environment inhibit transitions. Between these extremes lies a sweet spot: enough dephasing to break harmful interference and to blur mismatched energies, but not so much that all coherent mixing is strangled. That sweet spot is the "Goldilocks" regime: not too silent, not too noisy.

One can tell this story without equations, but it helps to carry one small piece of mathematical intuition in the pocket. Coherent dynamics are governed by energy differences and couplings. If two pigments (or two excitonic states) are far apart in energy, the coupling between them is effectively off-resonant; transfer is inefficient. Environmental fluctuations broaden energy levels and provide time-dependent shifts that can temporarily bring mismatched states into resonance. The environment is, in effect, continually trying different small retunings. In a narrow window of strengths and timescales, this dynamic disorder does not merely decohere; it also *matches*.

The key is timescale. Protein motions span a range: some are slow, like larger conformational changes; some are fast, like local vibrations. If the environment changes too slowly, it behaves like static disorder: each complex is slightly different, but within any given complex the mismatch remains, leaving transfer bottlenecked. If it changes too fast and too strongly, it randomises phases so quickly that coherent delocalisation cannot build, and the motion becomes overdamped. But if fluctuations have the right correlation time and amplitude, they act like a gentle tremor: enough to kick the exciton over small energy barriers without erasing every hint of structure.

A concrete example keeps this from feeling like wordplay. Imagine three pigments, labelled A, B, and C. Pigment C is closest to the reaction centre, so it is the one you want to populate. A and B are upstream. Suppose A couples strongly to B, and B couples to C, but the energy of B is slightly mismatched relative to C. In a perfectly coherent, perfectly static system, that mismatch can create oscillations between A and B while C remains underfed. Add a modest amount of environmental fluctuation that jitters B's energy. Now, at certain moments, B and C align better, and transfer to C becomes more likely. Add too much fluctuation and the system loses any coherent mixing that moved excitation from A to B efficiently in the first place. The best transport happens when the jitter is tuned to the mismatch.

This tuning language can tempt one into thinking biology must be exquisitely engineered for ENAQT, as though a protein has dialled in the perfect noise strength like an audio engineer setting reverb. The reality is subtler and, in its way, more plausible. Natural selection does not need to invent quantum mechanics; it only needs to avoid designs that waste energy. Pigment distances, orientations, and protein stiffness can all be nudged over evolutionary time. A complex that sits near the peak of the ENAQT curve will deliver more excitation to the reaction centre, improving fitness. There is no requirement that the system be tuned to the last decimal place. The peak is often broad: a range of noise strengths yields good performance. Robustness, rather than perfection, is the biological currency.

The Goldilocks picture also helps deflate a common misunderstanding: that more coherence is automatically better. Coherence can create speed, but it also creates patterns, and patterns can be misplaced. A fully coherent wave can form standing waves. In a musical instrument, standing waves are beautiful because they are stable and resonant. In an energy-transfer network, a standing wave that does not overlap the sink is disastrous. Dephasing breaks the symmetry that supports such a pattern, turning a stubborn standing wave into something that can leak into the exit channel.

There is a historical anecdote that captures the cultural whiplash this idea caused. In the first flush of excitement around ultrafast spectroscopy and "quantum beats," it was easy to picture biology as fighting the environment: coherence as a fragile gem that must be protected from thermal noise. The ENAQT models flipped the script. Suddenly, noise was not only the enemy; it could be part of the strategy. For researchers trained on the dream of isolating quantum systems, this was almost heretical. Some of the most productive conversations in quantum biology began with that discomfort: the real question became not "How does biology avoid noise?" but "What kind of noise does biology live with, and how does it shape what is possible?"

A practical way to anchor the Goldilocks effect is a simple energy-landscape cartoon. Energy transfer in these complexes is often "downhill" on average: excitations tend to move toward lower-energy sites that are closer to the reaction centre. But the downhill path is not smooth. There can be small barriers, local minima, and side valleys. Coherence alone can leave the excitation oscillating in a side valley; classical randomness alone can waste time wandering. The tuned tremor helps the excitation hop out of unhelpful basins without turning the motion into aimless diffusion.

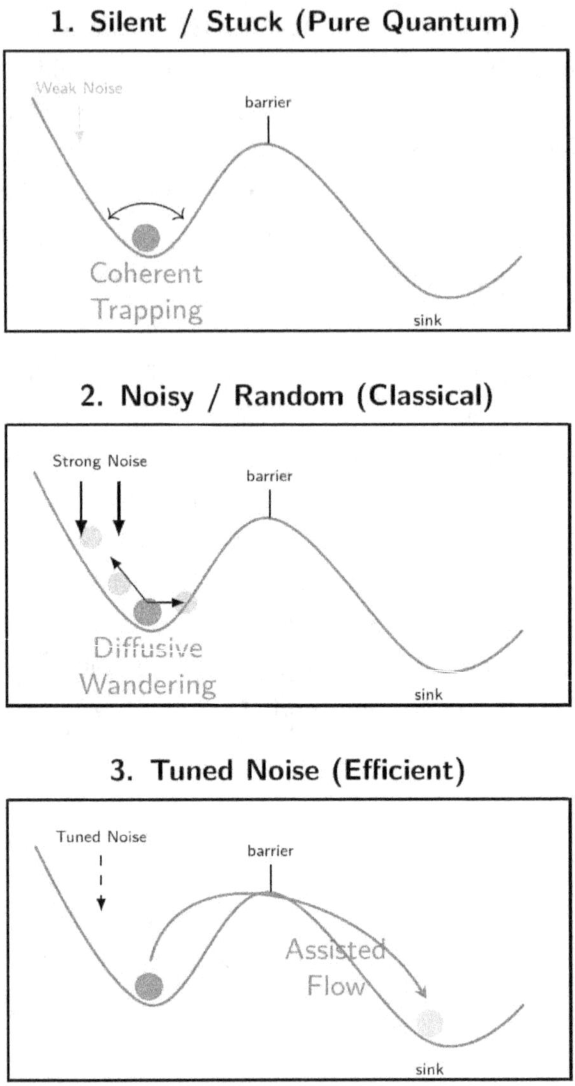

The diagram hides a key conceptual leap: the "ball" is not literally a marble rolling on a static hill. It is a quantum excitation whose motion is governed by amplitudes and phases. The "shaking" is not a hand physically rattling the landscape; it is the random, thermally driven fluctuation of pigment energies and couplings caused by molecular motion. Yet the analogy earns its keep because it captures what ENAQT insists on: the most efficient motion can require neither perfect determinism nor total chaos, but a controlled amount of disturbance.

One might wonder whether this is merely a trick of theory. Models can be made to do almost anything with enough parameters. The persuasive power of ENAQT comes from how often the qualitative behaviour appears across different network structures and modelling choices. Filippo Caruso and collaborators around 2009 emphasised that dephasing can suppress destructive interference and can also broaden energy levels, both of which can enhance transport. Alex Chin and colleagues added intuitive pictures in mixed site/exciton bases: noise can reshape pathways, weakening unproductive routes and strengthening more direct ones. Even when the details vary, the existence of a peak in efficiency versus noise strength is a recurring motif.

The caveats are real and scientifically healthy. The environment in a protein is not featureless white noise; it can have structure, with temporal correlations (memory) and spatial correlations (different pigments experiencing related fluctuations). Those structures can either help or hinder coherence and transport, depending on their character. Moreover, what spectroscopy calls "coherence" and what transport dynamics require are not always the same thing. A signal that oscillates beautifully in an experiment may correspond to coherences that are not strongly coupled to the sink. ENAQT is most compelling when it is linked to functional metrics: speed to the sink, yield, robustness against disorder, resilience when pigment energies vary.

The emotional heart of the Goldilocks shiver is its reversal of a familiar moral. Noise is usually treated as a thief of information. Here it can be a liberator. The protein's thermal motion, far from being an inconvenience, can play the role of a subtle assistant, loosening the grip of destructive interference and smoothing over energy mismatches. Life does not need to keep the quantum world in a glass case. It can let it rattle, slightly, and benefit from the rattle.

There is a final nuance that keeps the idea from turning into a slogan. The claim is not that noise *always* helps, or that biological systems must be precisely tuned to exploit it. The claim is that in coupled networks where quantum interference can create bottlenecks, a moderate degree of environmental interaction can increase transport efficiency by opening pathways that pure coherence would partially block. That is a statement about a physical mechanism, not a declaration of biological intention.

In the end, the exciton's best chance of reaching the reaction centre may depend on a kind of disciplined imperfection: enough coherence to explore, enough fluctuation to avoid getting stuck, and enough time pressure that the whole process finishes before energy is lost. Under the soft, incessant shiver of molecular life, the excitation can slip past barriers that a perfectly silent quantum world would ironically make impassable.

4.3 Ripples or Pebbles?

A wavy line on a graph can start an argument faster than a politics column. In the years after the first reports of long-lived oscillations in ultrafast spectra of photosynthetic complexes, the field found itself staring at the same experimental traces and telling two very different stories. One story heard a quantum ripple: a coherent exciton moving through a pigment network like a wave, sampling routes in parallel. The other heard pebbles: ordinary molecular vibrations, the whole protein-and-pigment assembly ringing mechanically after being struck by an ultrashort pulse of light. Both stories could produce "beats." Both could sound plausible at a dinner table. Only one could carry the weight of the bigger claim that biology makes deliberate use of quantum coherence.

The trouble begins with a deceptively simple fact: every molecule is a bundle of springs. Atoms in a pigment do not sit still; they stretch and bend in quantised vibrational modes. Even in the dark, those modes are thermally excited. When light excites the pigment electronically, it can also displace the vibrational potential surface, launching a wavepacket of nuclear motion—a coherent vibration that is as real, in its own way, as the vibration of a guitar string after a pluck. Ultrafast laser pulses are excellent plucks. They arrive so quickly that the nuclei cannot move during the excitation, and the molecule is left vibrating on the new electronic surface.

If that vibration stays phase-coherent for a while, it can imprint oscillations on optical signals. Those oscillations can persist even when electronic coherence has died, because nuclear motions can be more robust, or because their frequencies are sharply defined. A spectroscopist sees an oscillation at a certain frequency and must decide: is this the phase evolution between two electronic (excitonic) states, or is it the periodic motion of nuclei, or some hybrid of the two?

That third possibility—the hybrid—is where the debate becomes most interesting. A pigment is not simply "electronic" or "vibrational." Electronic excitation changes the forces on the nuclei; nuclear motion changes the electronic energies. When the coupling between an excitonic energy gap and a vibrational frequency is strong enough, the eigenstates of the combined system can mix. The result is a vibronic state: part electronic, part vibrational, neither purely one nor the other. In such a case, an oscillation in a spectrum could be driven by vibronic coherence, and the question "quantum or classical?" becomes ill-posed. The more useful question becomes: what kind of coherence is present, how long does it last under physiological conditions, and does it affect energy transfer to the reaction centre?

The historical anecdote that captures the mood of this period is almost theatrical. After the 2007-style results, seminar rooms filled with a new kind of excitement: photosynthesis as a stage for quantum mechanics in action. Titles began to contain words like "coherence" and "entanglement." Some talks leaned toward a bold narrative of quantum design; others were cautious, pointing out that light-harvesting complexes are strongly coupled to their environments, with broad spectral lines that usually signal fast dephasing. What seemed like a technical dispute over signal assignment carried an emotional charge because it touched a larger cultural desire: to find quantum weirdness not only in laboratories but in the heart of living systems.

The point of disagreement was not whether oscillations existed. They did. The point was what they meant. In 2D electronic spectroscopy, the emitted signal is a sum of contributions from different quantum pathways, often represented with Feynman-like diagrams. Some pathways involve coherence between excited states (an excitonic coherence). Others involve coherence between ground-state vibrational levels (a vibrational coherence that lives even when the electronic system is back in the ground state). Depending on pulse ordering and polarisation, the experiment can emphasise different pathways. This is why the debate became so entwined with experimental technique: changing the polarisation sequence is not mere refinement, but a way of asking a different question of the sample.

A useful mental picture is to imagine two different objects that both ring when struck. One is a pair of tuning forks of slightly different pitch. Strike them together and you hear beating—an amplitude modulation caused by superposing two nearby frequencies. That resembles excitonic beating: a superposition of two energy eigenstates evolving with two phase rates. The other object is a single bell with a rich set of vibrational modes. Strike it and you also hear oscillations, sometimes long-lived, because mechanical vibrations are underdamped. That resembles vibrational coherence. With only the sound, you might confuse them. In spectroscopy, the "sound" is the oscillatory component of a signal, and the job is to identify the instrument.

One line of evidence that shifted many opinions came from polarisation-controlled 2DES studies and careful modelling of the FMO complex. By selecting pulse polarisations that suppress certain contributions and enhance others, researchers could reduce signals associated with purely vibrational coherences and examine what remained. The emerging picture in much of this work was that the truly electronic coherences in FMO dephase quickly—on the order of a few hundred femtoseconds or less—whereas longer-lived oscillations in the spectra could be largely assigned to vibrational coherences or vibronic contributions. The beats were real, but many were not the rippling exciton-wave once imagined.

Temperature dependence offered another lever, as honest as it is brutal. If an oscillation corresponds to electronic coherence between excitonic states, it should be sensitive to thermal fluctuations that scramble electronic phases. If it corresponds to certain vibrational coherences, it may persist more robustly. Experiments that tracked oscillations across a range of temperatures found a trend that fits this intuition: as temperature rises, electronic coherence becomes harder to maintain, while some vibrational signatures remain. In FMO, work published years after the initial excitement suggested that above roughly 150 K, electronic coherence becomes difficult to argue as functionally significant, even if oscillatory signals can still be detected in certain spectroscopic channels. By the time one reaches physiological temperatures, the room for long-lived electronic coherence narrows considerably.

This could sound like a disappointment: a deflation of a romantic quantum story. Yet the debate sharpened something more valuable than romance: clarity about what would count as evidence of a functional quantum effect. A long-lived oscillation in a spectrum is not itself a functional claim. Function requires a link to energy transfer yield, speed, or robustness. It also requires confronting the nature of the excitation source. Sunlight is incoherent, arriving as a stochastic stream of photons with coherence times far shorter than the engineered pulses used in 2DES. A laser-prepared coherent superposition is an experimentally useful state, but it is not the state produced under natural illumination. This mismatch does not invalidate spectroscopy, but it restricts what one can infer.

A concrete example makes the restriction tangible. Suppose a 2DES experiment produces an excited-state coherence lasting 1 picosecond under cryogenic conditions, and that coherence produces beautiful beats. In nature, the complex absorbs sunlight, creating an initial state closer to an incoherent mixture across excitonic states. The long-lived coherence might never be launched in the same way. Even if it could be, the question is whether it changes the probability of reaching the reaction centre before decay. A coherence that is long-lived but confined to a part of the network that is not strongly connected to the sink might be spectroscopically prominent yet functionally irrelevant.

This is one reason the ENAQT idea sits awkwardly but fruitfully alongside the beats debate. ENAQT says that a certain amount of dephasing noise can help transport by preventing coherent trapping. That implies that maximal coherence is not necessarily desirable. So even if electronic coherence exists briefly, it need not be long-lived to matter. A flash of coherence that sets up efficient delocalisation early on, followed by noise-assisted relaxation toward the sink, could be enough. In that blended view, it would be a mistake to measure success by coherence lifetime alone. A short-lived ripple can still steer the flow before it dissolves.

The field's arguments also revealed a deeper conceptual trap: the temptation to sort everything neatly into "quantum" versus "classical." Vibrations are not automatically classical. Vibrational modes are quantised; they can be prepared in coherent states; they can, under certain conditions, show unmistakably non-classical features. The more precise question is operational: what signatures demonstrate genuinely quantum behaviour that cannot be mimicked by a classical stochastic model? Some theorists proposed criteria based on non-classicality in phase space or on correlations that violate classical bounds. These are hard to measure in complex biomolecules, but they serve as a reminder that "vibration" is not a synonym for "banal."

Artificial molecular systems have played an important role here, because they allow cleaner control. In certain designed aggregates and synthetic light-harvesters, vibronic coupling can be engineered so that a vibrational frequency resonates with an excitonic splitting. Under those conditions, long-lived beating in spectra can be observed with fewer biological complications, and the origin can be pinned down more convincingly. These systems provided something like wind-tunnel tests for ideas that were harder to isolate in natural complexes. They also taught a sobering lesson: resonance alone does not guarantee long-lived coherence. The environment's decoherence rates—how quickly different degrees of freedom lose phase information—can determine whether a beat is visible at all. You can tune a system to a perfect energetic match and still see no sustained oscillation if the environment damps it quickly.

That lesson fed back into the interpretation of biological data. Photosynthetic complexes are not designed for spectroscopic beauty. They are designed, in the only sense evolution designs anything, for reliable function under messy conditions: varying light, temperature changes, mutations, oxidative stress. A mechanism that depends on exquisitely long-lived electronic coherence at room temperature would be biologically fragile. A mechanism that uses a combination of short coherent dynamics, vibronic structure, and noise-assisted relaxation could be more plausible, because it is robust: it works even when coherence is not pristine.

The debate has also had a methodological payoff. It pushed experimentalists to craft cleverer controls: changing temperature, isotopic substitution (which shifts vibrational frequencies), modifying pulse sequences, and using polarisation tricks to separate contributions. It pushed theorists to build models that link spectroscopic observables to transport dynamics, rather than assuming they are the same. Yuta Fujihashi and Graham Fleming, among others, emphasised that long-lived oscillations in spectra can persist even when the transport-relevant electronic coherences are short-lived, because the spectroscopy can be sensitive to vibronic mixing that does not translate straightforwardly into improved energy transfer under natural conditions. That distinction is, in a way, the mature heart of the field: a recognition that seeing a quantum signature is not identical to proving a quantum advantage.

It is also worth acknowledging the social dynamics that any young interdisciplinary field must endure. Quantum physicists are trained to demand clean signatures and to distrust warm, wet systems; biochemists are trained to demand physiological relevance and to distrust overinterpretation. Ultrafast spectroscopists live in the middle, extracting meaningful information from signals that are beautiful but complicated. The "ripples or pebbles" argument was partly a collision of these cultures, with different instincts about what counts as convincing. The outcome has been less a verdict than a sharpening of shared standards.

For a reader trying to keep hold of the main thread, one can reduce the controversy to a simple question asked in two languages. In the language of physics: do the observed oscillations correspond primarily to electronic coherence between excitonic states, or to vibrational/vibronic coherences, and on what timescales do each survive at relevant temperatures? In the language of biology: does any of this measurably affect how efficiently excitation reaches the reaction centre under natural conditions? The first question is about assignment; the second is about function. The field has become increasingly comfortable with the possibility that the first answer can be "mostly vibrational at long times" while the second answer remains "transport is still exceptionally efficient," with ENAQT-like mechanisms helping to explain why.

A final image can hold the nuance without forcing a false choice. Imagine walking across a shallow stream on stepping stones. If the stones are placed well, you can cross quickly by hopping. Now imagine the same stream with a thin sheet of ice forming on top. The ice is not thick enough to bear your weight everywhere, but it can distribute pressure and smooth your path for a moment, allowing longer strides before it cracks. Neither picture is complete by itself. In photosynthetic complexes, some dynamics look like hopping, some like brief wave-like spreading, and some like the structured trembling of the molecular scaffold itself. The oscillations in spectra are the audible evidence that something is ringing. Determining whether it is a ripple in an excitonic wave, the ringing of nuclear pebbles, or a coupled harmony of both is less a hunt for a single magic mechanism than a careful sorting of timescales, couplings, and consequences.

When that sorting is done honestly, the most intriguing conclusion is not that biology is secretly performing laboratory-grade quantum tricks, but that the boundary between quantum coherence and classical motion in living matter is textured. In that texture, even a short-lived coherence can coexist with helpful noise, and even an apparently "classical" vibration can participate in the energy flow. The exciton does not need to remain a perfect wave to benefit from being one for a moment, and the protein does not need to be quiet to be helpful. In the crowded interior of a photosynthetic complex, a little ringing can be part of the way energy finds its way home.

Chapter 5

The Ghost in the Machine

Inside every cell, billions of chemical reactions occur at speeds that seemingly violate the standard laws of thermodynamics. This chapter investigates the mystery of enzyme efficiency, revealing how life employs a ghostly quantum shortcut-tunneling-to bypass energy barriers that would otherwise freeze biological time.

5.1 The Impossible Speed Limit

Life runs on chemistry, but not on chemistry as it behaves in a beaker left to its own devices. In plain water at room temperature, many reactions that cells rely on are not merely slow; they are so slow that a universe with only "ordinary" chemistry would be a quiet universe. The uncomfortable fact is that metabolism is full of steps that, outside the cell, would take longer than a human lifetime, longer than civilizations, sometimes longer than geological eras. Yet inside a bacterium or a liver cell these same transformations happen fast enough to keep pace with breathing, thinking, moving, growing.

The word that covers most of this miracle is *enzyme*. Enzymes are proteins (with a few RNA exceptions) that speed up chemical reactions without being used up. That bland definition hides the scandal.

In countless cases an enzyme accelerates a reaction by factors of millions, billions, and, in some celebrated comparisons, up to a trillion relative to the uncatalyzed process in water. A trillion is 10^{12}: the difference between a second and about 32,000 years. If that sort of claim were made about a sports car or a computer chip, you would look for the trick. In biology, the trick is built into the molecular architecture of the catalysts that cells manufacture by the billion.

A traditional way to picture enzymes is the "lock and key" metaphor associated with Emil Fischer in the 1890s: the enzyme is a lock machined to fit a particular molecular key, the substrate. The metaphor is useful as far as it goes. Many enzymes *are* selective; they distinguish among molecules that, to our eyes, look nearly identical. A digestive enzyme can attack one sugar but ignore its mirror-image. A detoxification enzyme can recognize a toxin at concentrations so low you could never taste it. The active site, a pocket formed by the folding of the protein, has a chemistry and a shape that lets it bind one substrate better than another.

But the lock-and-key picture carries a quiet assumption: once the key is in the lock, chemistry happens more or less the same way it would in solution, only guided into place. That assumption does not survive contact with the speed numbers.

To understand why, it helps to borrow a simple image from physical chemistry: the *energy landscape*. For two molecules to react, they typically have to pass through a high-energy, fleeting arrangement of atoms called the *transition state*. This is not an intermediate you can bottle; it is the crest of a hill that separates reactants from products. The height of that hill, in terms of free energy, is the *activation free energy*. At room temperature, thermal motion provides molecules with a jostling supply of energy. Most collisions are too gentle or too misaligned to reach the crest; a small fraction succeed, and that fraction sets the reaction rate.

A first-year chemistry story says: if you want a reaction to go faster, you either heat it up (give molecules more thermal energy) or provide a catalyst that lowers the hill. Life rarely has the option to heat things up; cells operate in narrow temperature windows. So enzymes must be in the business of hill-lowering.

That is true, but it is not the whole truth, and the details matter because "lowering the hill" is deceptively hard to do. The transition state is a particular pattern of stretched and partly broken bonds, along with charges rearranged in mid-flight. To lower the hill, the enzyme has to stabilize *that* strained arrangement more than it stabilizes the reactant. It is as if the lock were built not to fit the key you start with, but to fit a key you only have for an instant while you are turning it. This idea, developed into a mature theory in the mid-twentieth century and associated with Linus Pauling, is often summarized as "transition-state stabilization": the enzyme's active site is complementary in shape and electrostatics to the transition state, and by binding it strongly the enzyme reduces the activation free energy.

Even with that more sophisticated picture, the raw scale of enzymatic acceleration still feels like an insult to intuition. Consider what "a trillion times faster" means in energetic terms. Reaction rates depend exponentially on activation free energy, roughly as $\exp(-\Delta G^{\ddagger}/RT)$, where $\Delta G^{\ddagger}$ is the activation free energy, R is the gas constant, and T is temperature. To speed a reaction up by a factor of 10^{12} at room temperature, you do not need to eliminate the barrier entirely, but you do need to reduce the activation free energy by on the order of 16—17 kcal/mol. That is not a tiny tweak. It is comparable to the energy of several strong hydrogen bonds, arranged with exquisite precision, and it has to be delivered to an arrangement of atoms that exists for a fraction of a trillionth of a second.

So the first "speed limit" is energetic and probabilistic: thermal energy is not a controlled resource. It is messy. A water molecule hits the substrate here, a sodium ion drifts by there. If the only tool available were random thermal agitation, you would expect the reaction to proceed at whatever pace the environment happens to allow, with modest improvements if a molecule holds two reactants close together. Yet enzymes routinely do far more than modest improvements.

The second "speed limit" is geometric. Chemical reactions are not only about having enough energy; they are about meeting in the right orientation. If you have ever tried to insert a key into a lock in the dark, you have experienced the tyranny of alignment. Molecules in solution tumble and collide in all orientations. A catalyst that merely brings partners together, without arranging them precisely, does not escape the tyranny; it just increases how often collisions happen. There is a ceiling to this kind of acceleration: even if a substrate molecule were perfectly reactive upon collision, it can only collide so many times per second. This is the realm of *diffusion control*. Many enzymes that have evolved to deal with toxic compounds or to process abundant substrates approach this ceiling; they are sometimes called "catalytically perfect," not because they cannot be improved in principle, but because they are nearly as fast as the physics of molecular wandering in water allows.

Diffusion control is a revealing limit because it separates two kinds of "fast." There is fast because the chemistry itself has been made easy, and there is fast because the chemistry was never the bottleneck to begin with. When an enzyme is diffusion-limited, the slow step is not the bond-breaking event; it is simply the time it takes the substrate to bump into the enzyme and enter the active site. That may sound like an uninteresting detail, but it is an important clue: in such cases the enzyme has pushed the chemical step to be so efficient that the environment's randomness becomes the constraint.

But many metabolic steps are not diffusion-limited. They involve transformations that are intrinsically difficult: splitting stable bonds, transferring charged groups in water, reconfiguring electron distributions. For these reactions, the chemistry *is* the bottleneck in solution, and enzymes somehow make it easy enough to run on biological timescales.

A concrete example brings the point down to earth. Carbon dioxide is a small, unassuming molecule that your cells produce constantly. In water, carbon dioxide can slowly react to form bicarbonate, a reaction that matters for blood pH and for the transport of carbon dioxide out of tissues. Left alone, this hydration reaction happens, but not quickly enough for an animal that is sprinting or thinking hard. The enzyme carbonic anhydrase solves the problem so thoroughly that, under physiological conditions, it can process on the order of a million molecules per second per enzyme molecule. It is among the fastest known enzymes, and it is so efficient that in many regimes it brushes up against diffusion control: the enzyme spends much of its time simply waiting for the next molecule of carbon dioxide to arrive.

Carbonic anhydrase is a useful case because it makes it tempting to think that enzymatic speed is just a matter of "super-sticky binding" or "super-crowded collisions." Yet carbonic anhydrase does something more clever: it uses a zinc ion in the active site to activate water, generating a hydroxide that is a much better nucleophile than ordinary water. It arranges proton transfers through a network of residues and waters, providing a pathway that water in bulk solution cannot easily organize. The enzyme is less like a passive lock and more like a small machine: it positions, polarizes, and shuttles charges in a choreographed sequence.

That choreography hints at why classical "spheres bumping" pictures can be misleading. At the molecular scale, atoms are not billiard balls. Bonds are not rigid sticks. Electron clouds distort. Partial charges shift. Hydrogen bonds form and break. A protein is not a statue; it fluctuates, breathes, and samples many conformations, even at constant temperature. If one insists on a naive classical picture, the enzyme seems to require impossibly precise timing: the right configuration, the right alignment, the right charge distribution, all at once, every time, at extraordinary rates.

Chemists learned early that this precision is not a matter of luck. The enzyme's structure biases the fluctuations so that productive configurations are visited far more often than they would be in solution. The enzyme can bind substrates in a way that strains certain bonds, bringing them closer to the transition state. It can create microenvironments that are unlike bulk water, excluding water molecules that would stabilize the wrong charges and allowing others that stabilize the right ones. It can offer acid and base groups (amino acid side chains) at just the right positions to donate or accept protons. It can use metal ions, like zinc, magnesium, or iron, to stabilize negative charge or to orchestrate electron transfer. All of these are classical ingredients, and they are real.

And yet the "impossible" feeling returns when one looks at the most dramatic rate enhancements, especially in reactions involving hydrogen transfer. Hydrogen is the lightest atom; its nucleus is a single proton. Many central metabolic steps require moving that proton, or moving a hydrogen as a whole (a proton plus an electron, called a hydrogen atom transfer), or moving a hydride (a proton plus two electrons, effectively a H^- transfer). These processes lie at the heart of oxidation-reduction chemistry, the kind cells use to extract energy from food and to build complex molecules.

Here is the classical obstacle. Imagine a hydrogen being transferred from a donor atom to an acceptor atom. Classical transition-state theory encourages a picture in which the hydrogen sits on the donor, thermal motion rattles the system, and occasionally the hydrogen gains enough energy to crest the barrier and land on the acceptor. The barrier is not only about energy; it also has width. The hydrogen must pass through a region where it is bonded to neither atom in a stable way, and where electrostatics are in awkward transition.

In many enzyme active sites, the donor and acceptor atoms can be brought close, and the electrostatic environment can be tuned. That lowers and narrows the barrier. But there is still a mismatch between what thermal energy seems able to provide and the observed rate in some systems, especially when the reaction in water is excruciatingly slow and the enzyme-driven rate is astonishingly fast.

Richard Wolfenden, who spent decades measuring reaction rates that are so slow they test one's patience and instrument design, helped sharpen this contrast by comparing enzyme-catalyzed reactions with their uncatalyzed counterparts under carefully controlled conditions. The point was not that enzymes make every reaction faster by the same factor. The point was that for certain transformations, water is a remarkably poor catalyst and the uncatalyzed rate can be almost unimaginably low. If an enzyme then performs that chemistry in milliseconds, it looks less like an incremental improvement and more like a different physical regime.

At this stage, it is easy to overreact in the other direction and declare classical chemistry defeated. But it is more instructive to identify exactly where the classical picture begins to creak. A good place to watch it creak is in the notion that rate enhancement is always explained by a single, static reduction in activation free energy: the enzyme "lowers the barrier," end of story. Real enzymes do lower barriers, but they do so in environments that are constantly fluctuating. The barrier is not a fixed hill; it is a living landscape whose shape changes as the protein moves. Some conformations are catalytically poised, aligning donor and acceptor, placing charges, and shaping a narrow passage; others are less helpful. The enzyme can spend time searching its own conformational space, and the chemistry can be sensitive to rare but crucial "reactive" substates.

That dynamic view remains compatible with classical physics, but it carries an implication: in many cases the catalytic event depends on brief alignments that are better described as changes in *barrier width* and *barrier shape* than simply barrier height. And the moment one talks about barrier width for a very light particle like a proton, quantum mechanics starts whispering from the margins.

One might object: quantum mechanics is always present at the molecular scale; why should enzymes be special? The electrons that make chemical bonds are quantum objects everywhere, whether in an enzyme or in a flask. That is true, and it is precisely why quantum biology must be careful not to claim magic where none is needed. The interesting question is not "Is chemistry quantum?" It is "Are there quantum effects that are *functionally* significant for how biology works under warm, wet conditions?" In enzymes, the most plausible candidates involve the nuclei of hydrogen atoms, not because enzymes turn off classical physics, but because hydrogen's small mass makes its quantum nature harder to ignore.

To keep this grounded, return to the speed limit in its simplest form. A reaction with a large activation free energy is slow because only a tiny fraction of molecular configurations have enough energy at any instant to reach the transition state. Enzymes speed reactions by stabilizing the transition state and by organizing the environment so that productive configurations are not astronomically rare. Classical transition-state theory packages this into a tidy formula, but the tidy formula assumes a smooth passage over the barrier, like a ball rolling over a hill. For hydrogen transfer, especially when the donor and acceptor are not extremely close, the barrier may be too wide for the ball picture to be the most faithful one. The hydrogen can behave less like a localized particle waiting to be kicked, and more like a smeared-out wave with a nonzero chance of appearing on the other side even when it does not have the classical energy to climb over.

This is where the "impossible" character of enzymatic speed becomes useful rather than frustrating. It forces a more precise question: what, exactly, is the enzyme optimizing? If it were only raising the substrate concentration near reactive residues, the gains would be limited. If it were only holding the substrate in place, the gains would be substantial but not always trillion-fold. If it were only lowering a barrier height, then the remaining rate should match what you would predict from the new height, and the details of mass and temperature dependence should behave in a particular semiclassical way.

Instead, enzymologists began to notice patterns that do not sit comfortably inside a purely over-the-top picture of thermal activation. A telling feature is that many of the most dramatic enhancements occur in steps that involve hydrogen transfer, and that subtle changes in the distance between donor and acceptor atoms can have outsized effects on rate. Distance is not merely geometry; it is probability, because it shapes the barrier width. In a warm protein, distances are not fixed. Side chains wag. Hydrogen bonds stretch and compress. Water molecules enter and leave. A protein can, through its fluctuating structure, repeatedly sample configurations where the donor and acceptor draw close enough for the hydrogen transfer to become exceptionally likely.

The older lock-and-key metaphor becomes inadequate here, not because it is wrong, but because it is static. A more faithful image is a glove that flexes and tightens at the right moment, briefly creating a fit that does not exist in its relaxed state. Daniel Koshland's induced-fit concept, introduced in the late 1950s, made this dynamism explicit: binding and catalysis can reshape the enzyme, and that reshaping can be part of the catalytic mechanism itself. In practice, enzymes use both preorganized features (a pocket already predisposed to stabilize a transition state) and induced features (movements triggered by binding) to reach their most reactive configurations.

There is also a social history to this, because the numbers were so extreme that they demanded better measurements. When enzymologists and physical chemists began comparing catalyzed and uncatalyzed rates systematically, they were not only collecting impressive facts; they were testing the idea that biological catalysis could be reduced to the same principles that govern catalysis in simpler chemical systems. Some of the most persuasive experiments were conceptually simple: measure the same reaction with and without the enzyme under comparable conditions, and compare. But "without the enzyme" can mean a rate so slow that it is hard to measure directly, forcing clever approaches: high temperatures extrapolated down, alternative detection methods, or indirect calculations grounded in thermodynamics.

The moral is not that enzymes break the rules. The moral is that the rules, as commonly told, hide the resources that matter. The rules of classical transition-state theory, in their popular form, are often taught with the tacit assumption that the barrier is fixed and the reactants are point particles that must climb it. In a living enzyme, the barrier is sculpted by a fluctuating environment that can compress distances and reshape electrostatic fields. When the moving piece is hydrogen, the distinction between a barrier that is high and a barrier that is wide becomes more than semantics.

One can feel this in a thought experiment. Imagine a reaction barrier as a stone wall. Classical thermal activation is like giving a runner enough energy to sprint over it. Lowering the wall helps; giving the runner a better run-up helps; positioning the runner exactly helps. But if the wall is thick, even a wall that is not terribly high can be hard to clear cleanly. Now replace the runner with a mist. A mist does not “jump” in the same way. Some of it can seep through cracks you did not know were there. If you can narrow the wall, even without lowering it much, the seepage can become significant. Enzymes, in many hydrogen-transfer reactions, appear to be expert wall-narrowers.

That is not yet a claim about quantum mechanics as the main driver of all enzymatic speed. Many enzymes are fast for reasons that are overwhelmingly classical: they stabilize charges, organize water, and use acids, bases, and metals in ingenious arrangements. Even when hydrogen tunneling contributes, it may contribute to the *chemical step* without being the dominant reason the enzyme outperforms water by ten or twelve orders of magnitude; the majority of the enhancement can still come from transition-state stabilization and electrostatic preorganization. The point, rather, is that the extremity of enzymatic rate enhancements opens a door: once the chemistry has been optimized so fiercely that the remaining bottleneck lies in subtle geometries and fleeting configurations, effects that would be negligible in a sloppier environment can become decisive.

It is also worth noticing that biology cares less about philosophical purity than about throughput. A cell does not need a catalyst that is elegant; it needs one that is reliable at 37°C in salty water amid a crowd of other macromolecules. Evolution has no reason to avoid using quantum-mechanical allowances if they are available. The real constraint is not whether an effect sounds exotic, but whether it survives in the noisy thermal bath of the cell. If an enzyme can, by shaping distances and vibrations, make a light nucleus behave in a way that boosts the odds of the crucial bond rearrangement, that is the kind of advantage natural selection can repeatedly seize.

The “impossible speed limit,” then, is not an official constant of nature. It is a pressure point in our understanding: the place where a simplified classical story about barriers and bumps begins to feel too thin for what enzymes actually do. Enzymes are not just passive holders of substrates. They are dynamic electrostatic devices that reshape the landscape a reaction moves across, sometimes making the decisive step less like climbing and more like finding a hidden passage. In the tightest corners of metabolism, where a single hydrogen must move at exactly the right time and place, the passage can be narrow enough that the ordinary picture of thermal surmounting stops feeling like the whole explanation, and the physics of very small masses starts to matter.

5.2 The Vanishing Proton

A useful way to think about enzymes is that they are experts at arranging circumstances. They crowd the right atoms into the same cramped room, point them in the right directions, and hold them there long enough for chemistry to happen. That story already explains a lot. But when the crucial moving piece is hydrogen, something subtler begins to matter: the hydrogen nucleus is so light that it does not behave like a tiny billiard ball waiting for a kick. It behaves, in part, like a spread-out *possibility*.

Chemists call hydrogen transfer by many names, depending on what is moving. Sometimes it is a proton, H^+, a bare nucleus; sometimes it is a hydride, effectively H^-, carrying two electrons; sometimes it is a hydrogen atom, $H\cdot$, moving together with one electron. Living cells use all of them. Acids and bases shuffle protons as casually as coins. Metabolic workhorses called dehydrogenases move hydrides between molecules, controlling the flow of electrons that ultimately powers muscles and thoughts. Many reactions are hybrids: a proton move is tied to an electron move in what is called *proton-coupled electron transfer*. The labels are bookkeeping. The deeper point is that hydrogen is constantly being relocated in biology, and those relocations often determine the pace of entire pathways.

In everyday life, barriers are crossed by going over them. If you lack the energy to climb a wall, you bounce off. Classical transition-state theory shares that instinct. The reactant must reach the transition state, the top of the hill, and the rate depends on how often thermal motion supplies enough energy. A proton transfer, in that picture, is a proton that sits on one side of a barrier and waits for a shove.

Quantum mechanics keeps the hill but changes what it means to be "on one side." At microscopic scales, particles such as electrons and protons also behave as waves. That is not a poetic analogy; it is the operational content of the mathematics. A wave is not located at a single point. It is distributed across space, with an amplitude whose square gives a probability of finding the particle at different positions if you measure it. For an object as heavy as a baseball, the wave is so absurdly localized that classical thinking works. For a proton, especially in a confined molecular environment, the wave can have a meaningful breadth.

Now picture the barrier again. In quantum mechanics, a particle's wave does not necessarily stop where classical motion would stop. It can leak into classically forbidden regions. If the barrier is thin enough, a small part of the wave can leak all the way through, creating a nonzero probability that the particle will be found on the far side without ever having had the classical energy needed to crest the hill. This is *quantum tunneling*. The word is apt because, in the mathematical description, the particle behaves as if it had passed through the barrier rather than over it.

The effect is often introduced with a melodramatic example: a particle crosses a wall it cannot possibly climb. It sounds like a magic trick until one remembers that "cannot possibly" is a classical statement about a particle with a definite position and a definite energy. A quantum particle does not carry those definite values in the same way between measurements. The wave description is not a convenient ignorance; it is the best possible account of what can be predicted.

Tunneling is not rare in physics. It is how some radioactive nuclei decay. It is how the scanning tunneling microscope sees atoms on surfaces. It is how the Sun can fuse hydrogen at temperatures that, by naive classical calculations, seem too low for protons to overcome their mutual electrostatic repulsion. The surprise in biology is not that tunneling exists, but that it can matter inside proteins at room temperature, in water, in a noisy environment where delicate quantum behavior is often assumed to be washed out.

Hydrogen's small mass makes it the prime suspect. In the crude but useful intuitions of quantum mechanics, lighter particles have "wavier" behavior: their wavefunctions spread more and, for a given barrier, their tunneling probabilities are higher. The difference between a proton and a heavier atom is not a matter of taste; it shows up quantitatively. Swap hydrogen for something heavier and tunneling can collapse. This sensitivity to mass will later become a crucial experimental lever. For now, it provides a reason to treat proton motion as a special case.

There is also a geometrical reason tunneling becomes plausible in enzymes. The probability of tunneling depends steeply on barrier width. A thin barrier is more transparent to a wave than a thick one. Enzymes, by compressing the donor and acceptor atoms and by shaping the surrounding electrostatics, can change both barrier height and barrier width. And unlike the bulk solution, where donor and acceptor positions are set by fleeting collisions, an enzyme can hold them close, sometimes with angstrom-level precision (an angstrom is 10^{-10} meters). It can also *fluctuate* them: slight breathing motions can repeatedly bring the key atoms into especially favorable proximity for a brief moment. Tunneling is not something an enzyme "turns on" like a switch; it is a probability that becomes meaningful when the barrier becomes narrow enough often enough.

A mental image helps, as long as it is kept on a short leash. Imagine a proton in an enzyme active site, sitting near a donor atom, with an acceptor atom a short distance away. Between them lies an energetic barrier: perhaps the acceptor is not yet charged appropriately; perhaps the geometry is not yet ideal; perhaps the surrounding water and side chains are arranged in a way that resists the transfer. In a classical picture, the proton waits until random thermal agitation gives it enough energy to go over. In a quantum picture, the proton is described by a probability wave. Most of that wave is still concentrated near the donor. But a tail extends toward the acceptor and can penetrate the barrier region. If the acceptor comes close enough, or if the barrier narrows during a protein fluctuation, the tail can reach the far side. At that point, the proton has a chance of "appearing" on the acceptor side. From the standpoint of a chemical observer who cares about rates, it looks as if the proton has crossed without climbing.

This is the core schematic idea:

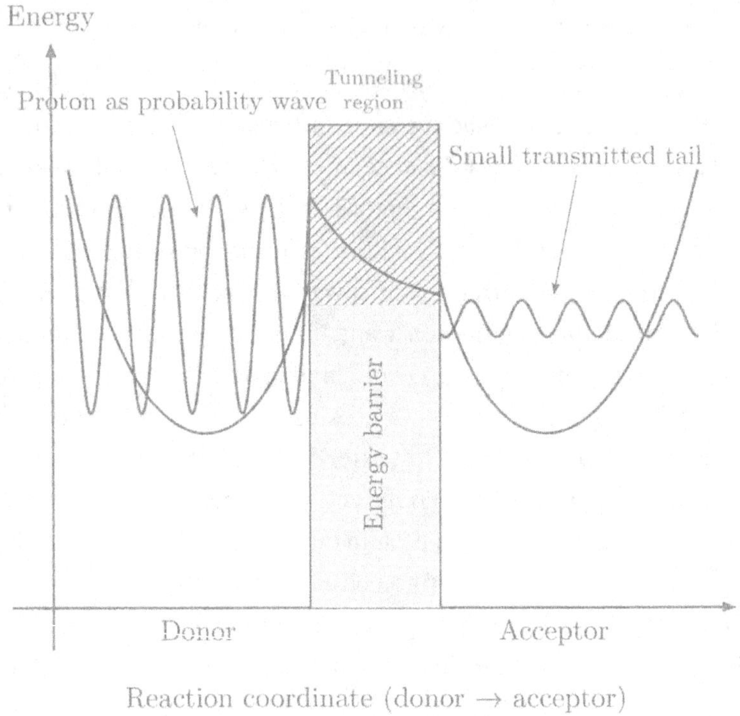

The point of drawing the proton as "fuzzy" is not artistic license; it is a statement about what the theory says is physically meaningful. The proton is not smeared in the sense of being physically chopped into pieces; it is smeared in the sense that, before a measurement pins it down, the best description assigns probabilities to where it might be.

Tunneling also invites a different way of thinking about time. In a classical picture, one can imagine the proton accelerating up the barrier, slowing at the top, then rolling down. In tunneling, that continuous trajectory is not the right mental model. What matters is coupling between quantum states: a state localized more on the donor side and a state localized more on the acceptor side. The barrier influences how strongly those states communicate. Narrow the barrier and the communication strengthens; the transfer can become fast even if the thermal energy does not change.

If this were all, one might worry that tunneling in enzymes is a seductive idea in search of a necessity. But tunneling's fingerprints are distinct enough that chemists have found it difficult to ignore. Long before "quantum biology" became a phrase that could sell a lecture, enzymologists were noticing oddities in reactions that move hydrogen.

A particularly instructive historical thread runs through alcohol dehydrogenase, an enzyme that catalyzes the oxidation of alcohols to aldehydes or ketones. In yeast, alcohol dehydrogenase plays a central role in fermentation, helping cells manage their electron bookkeeping when oxygen is scarce. The chemical heart of the reaction is the transfer of a hydride from the alcohol to a cofactor called NAD^+, producing NADH. That hydride transfer is a hydrogen motion tied intimately to electron rearrangement.

In the late 1980s, experiments using different hydrogen isotopes in yeast alcohol dehydrogenase, including tritium (hydrogen with two extra neutrons), produced kinetic patterns that did not line up neatly with semiclassical expectations. If hydrogen transfer were purely an over-the-barrier thermal event, certain relationships between isotope effects would be expected. Deviations suggested that the mass of hydrogen was influencing not only the energetic bookkeeping but the manner in which the particle crossed the barrier. The details require some technical definitions, but the emotional punch is simple: changing the mass of the moving hydrogen changed the reaction more than classical intuition said it should, and it did so even at ordinary temperatures, between 0°C and 40°C.

The natural skeptic's reply is that enzymes are complicated. Perhaps the isotope subtly changes binding. Perhaps it changes the vibrational modes in a way that shifts the transition state. Perhaps the supposed tunneling signature is just classical chemistry in a complicated outfit. Those are sensible objections, and they have kept the field honest. But the suspicion of tunneling persisted because the "complicated classical" explanations began to look increasingly contrived compared to a straightforward quantum one.

There is another, even more human piece of the story. Many people's first encounter with tunneling comes from electronics, where it sounds clean and engineered. Biology feels messy by comparison. Yet the evidence for enzymatic tunneling grew not from romantic theories but from stubborn experimentalists confronting anomalies. Judith Klinman and collaborators, beginning in the 1980s and continuing for decades, helped establish the idea that hydrogen transfer in enzymes can involve significant nuclear quantum effects under physiological conditions. The way it was done was almost detective work: change one thing at a time, measure carefully, and see whether the patterns match the signatures predicted by different mechanisms.

Tunneling, in this biological setting, is not a lone proton ghosting through an impenetrable wall while the rest of the enzyme stands still. Proteins are warm and restless. They constantly jiggle, and those motions can aid tunneling by shaping the barrier. The most productive way to combine classical and quantum ideas is to imagine a hierarchy of motions. There are slower, larger-scale rearrangements of the protein that sample different conformations, and there are faster local motions that can compress the donor and acceptor distance. When, by chance, the enzyme visits a conformation that makes the barrier narrow, tunneling probability rises sharply. A tiny change in distance can yield a large change in rate, precisely because tunneling depends exponentially on barrier width.

This helps explain why tunneling in enzymes does not contradict the warm, wet, noisy cell. The enzyme does not need to maintain delicate quantum coherence across its whole structure. It needs only to exploit a local quantum event during a very brief window, in a specific coordinate involving hydrogen. The environment can be "classical" in many respects while still allowing this one quantum shortcut.

An apparent counterargument is temperature. Many people first meet tunneling in contexts that become obvious only at low temperatures, where thermal motion is suppressed and classical over-the-barrier pathways shut down. If tunneling in enzymes were merely a cold trick, it would not help biology much. But experiments have shown tunneling signatures persisting at elevated temperatures in certain systems, including thermophilic enzymes operating around 65°C. Even more intriguingly, temperature trends can be counterintuitive if one expects tunneling to simply take over as temperature drops. In some cases, cooling a thermophilic enzyme reduces catalysis in a way consistent with the idea that protein motions are needed to *enable* efficient tunneling by sampling and compressing the right geometries. Thermal motion, in other words, is not merely the shove that tries to push hydrogen over the barrier; it is also the dance that intermittently reshapes the barrier to make tunneling more likely.

At this point, a careful reader may ask: if tunneling is a quantum way of crossing barriers without the right energy, does it mean enzymes are cheating energy conservation? They are not. Energy is conserved. The enzyme is not creating energy out of nothing; it is changing which pathways are probabilistically accessible. In classical transition-state theory, the key improbability comes from the need to concentrate enough energy in the right coordinate at the right time. In tunneling, the improbability can shift toward the need to achieve a sufficiently narrow barrier geometry, which is a structural and dynamical problem. The enzyme can influence that structural problem strongly by its architecture.

Nor does tunneling mean that the proton literally "teleports" with no relation to the environment. The environment still matters intensely: electrostatic fields, hydrogen-bond networks, and solvent organization shape the potential energy surface the proton experiences. Tunneling occurs *through* that landscape, not outside it.

One might also wonder whether tunneling can be more than a curiosity: can it matter for life in a way that touches experience? It can, if it affects reaction rates enough to influence physiology. A small shift in the rate of a key hydrogen transfer step can change the throughput of a metabolic pathway, the ability of an organism to detoxify a compound, or the rate at which a cell repairs oxidative damage. Biology often lives on margins: a bacterium competing for a nutrient in a puddle, a neuron trying to keep up with rapid firing, a liver processing a burst of alcohol. In those contexts, a catalytic advantage does not need to be metaphysically large to be evolutionarily decisive. But tunneling can be large: it can change a rate by orders of magnitude when the barrier is at the edge of transparency.

It is tempting, once tunneling is on the table, to treat it as the universal explanation for enzymatic prowess. That temptation should be resisted. Enzymes are fast for many reasons, most of them classical and beautifully chemical: electrostatic preorganization, precise positioning, acid-base catalysis, metal cofactors, and the harnessing of binding energy to stabilize transition states. Tunneling belongs in a more specific category. It matters most in hydrogen transfer, and even there it is often entwined with classical protein motions that bring donor and acceptor into a tunneling-friendly configuration. The ghost is not a disembodied spirit; it is a quantum probability wave taking advantage of a momentarily favorable geometry.

A proton transfer in an enzyme can therefore be imagined not as a single heroic leap but as a quiet statistical event: countless fleeting attempts, most unproductive, punctuated by rare instants when the barrier becomes thin and the proton's wave has enough reach. In those instants the proton does not have to wait for a classical shove big enough to get over the top. It simply has to be given a wall thin enough to leak through.

Once that image has settled in the mind, a familiar object looks different. A protein is no longer just a scaffold that holds reactants; it is a fluctuating piece of matter that can sculpt probabilities. It does not abolish the hill. It reshapes the hill's geometry until a particle as light as hydrogen can do what waves do best: find a path where a point-like classical object would find only an obstacle.

5.3 Weighing the Evidence

A daring idea in science lives or dies by the quality of its measurements. Quantum tunneling, applied to biology, is daring not because tunneling is exotic in physics, but because enzymes are warm, wet, and busy. They are the opposite of the carefully isolated systems that introductory quantum mechanics trains us to trust. If tunneling genuinely contributes to enzymatic chemistry, it should leave fingerprints that cannot be explained away as mere cleverness in classical catalysis.

The most powerful fingerprint turned out to be almost laughably simple: replace hydrogen with a heavier version of itself and see what happens.

Hydrogen comes in isotopic varieties. Ordinary hydrogen has a nucleus consisting of a single proton. Deuterium has the same single proton plus a neutron, roughly doubling the mass of the nucleus. Tritium adds a second neutron, tripling the mass, and is radioactive, which makes it useful but also demanding to handle. Chemically, these isotopes are almost the same: they have the same electron structure, and chemistry is mostly electron business. If two molecules differ only by whether a particular position carries H or D, they look nearly identical to the enzyme's binding pocket.

Yet reactions care about mass in one critical way: mass shapes motion. Heavier nuclei vibrate more sluggishly in bonds; their zero-point motion (the residual quantum jiggle that never quite goes away) is smaller; their probability waves, in the tunneling sense, are less adventurous. That means hydrogen isotopes give a uniquely sharp test. If the rate-limiting step involves moving a hydrogen nucleus through a barrier, changing the mass should change the rate. If the step is purely classical over-the-barrier activation, the change should be modest and should follow well-behaved semiclassical expectations. If the step has a significant tunneling component, the change can be much larger and can behave in ways that classical transition-state theory struggles to mimic.

The technical term for this is the *kinetic isotope effect*, usually abbreviated KIE. It is simply a ratio of rates: how fast the reaction is with hydrogen, divided by how fast it is with deuterium (or tritium), with everything else held as constant as possible. A KIE of 2 means hydrogen is twice as fast as deuterium; a KIE of 10 means ten times faster. These are not abstract numbers. If an enzyme is just barely keeping up with the metabolic demands of a cell, a tenfold slowdown would be catastrophic.

In classical chemistry, primary hydrogen/deuterium KIEs (where the isotope substitution is made on the bond that is actually being broken or formed) often fall into a fairly narrow range, commonly somewhere below about 7 at room temperature for simple over-the-barrier models. There is nuance here: the expected value depends on details, and unusually large classical KIEs can appear in special circumstances. But as a rough rule of thumb, a KIE that soars far beyond the typical semiclassical range demands an explanation.

The field's early drama came from exactly this sort of demand. Alcohol dehydrogenase, an enzyme central to fermentation and to alcohol metabolism, provided one arena. In the late 1980s, isotope experiments using hydrogen, deuterium, and tritium in yeast alcohol dehydrogenase produced patterns that did not sit comfortably within semiclassical relationships among H/T and D/T effects over ordinary temperature ranges. The puzzle was not merely that the reaction slowed with heavier isotopes (that is expected), but that the *way* it slowed, and how the ratios behaved with temperature, suggested that the hydrogen nucleus was not simply cresting a barrier in the classical manner. For those who were already suspicious that hydrogen might tunnel in enzymes, the results felt like a confession.

Still, extraordinary claims require more than one compelling case. Enzymes are complicated, and any single enzyme could have an idiosyncratic mechanism. The case for tunneling became more persuasive when similar isotope fingerprints appeared across very different enzymes that all shared one feature: a chemically decisive step involving hydrogen transfer.

Judith Klinman's laboratory, beginning in the 1980s and extending through decades of work, played a central role in turning isotope effects into a discipline rather than a curiosity. The strategy was not to chant "tunneling" whenever a KIE looked large, but to build a coherent picture across multiple experimental axes: magnitude of the KIE, its temperature dependence, the effects of mutations that subtly change active-site geometry, and the way rates respond to changes in solvent and viscosity. The underlying reasoning is straightforward: if the enzyme is merely lowering a classical activation barrier, isotope effects should behave like a modest correction; if the enzyme is repeatedly sampling configurations that allow hydrogen to tunnel efficiently, isotopes should act like a blunt instrument that changes tunneling probability itself.

Temperature is a particularly revealing lever. Classical transition-state theory, in its simplest form, predicts an Arrhenius-like dependence: rates speed up with temperature because thermal energy helps systems reach the transition state. Isotope substitution changes vibrational frequencies and zero-point energies, so it can change the apparent activation parameters, but it does so in constrained ways. Tunneling can scramble these constraints. One widely discussed signature is an unusual relationship between Arrhenius parameters for different isotopes, sometimes even producing inverse relationships in pre-exponential factors. The details can feel forbidding, but the idea is not: if two isotopes appear to have the "wrong" temperature dependence relative to one another, something beyond a simple over-the-hill picture is at work.

A concrete example that became almost iconic is soybean lipoxygenase, an enzyme that catalyzes a reaction involving hydrogen-atom abstraction from a fatty acid. Here the moving particle is not merely a proton; it is effectively a hydrogen atom transfer bound up with electron rearrangement. Measurements in the early 2000s revealed extraordinarily large hydrogen/deuterium KIEs in the wild-type enzyme near room temperature, around eighty-fold in some reports, with surprisingly weak temperature dependence. Those two features together are hard to square with a simple classical barrier-crossing story. A large, weakly temperature-dependent isotope effect fits naturally with a picture in which the hydrogen is tunneling through a barrier that is kept narrow by the enzyme's architecture, so that temperature is not primarily paying the energetic cost of climbing; it is largely paying the cost of sampling a tunneling-ready geometry.

Mutations sharpened the argument further. Change a residue that slightly alters the packing around the active site, or that changes the flexibility that allows the donor—acceptor to approach, and the KIE can become even larger while the overall catalytic rate drops. That combination is telling. If the enzyme becomes worse at achieving a close donor—acceptor distance, the barrier effectively widens: overall rate falls because tunneling probability falls, and the difference between hydrogen and deuterium becomes even more dramatic because the heavier isotope is even more sensitive to the widened barrier. In some double mutants of soybean lipoxygenase, KIEs reportedly climbed into the hundreds. That is not the sort of number one expects from a minor classical vibrational tweak.

One can summarize the logic without hiding the caveats. A KIE by itself is not a proof of tunneling, because chemistry is inventive and can produce large isotope effects by other mechanisms. But the *pattern* of large KIEs, their temperature dependence, and their response to mutations that plausibly change donor—acceptor distance provides a triangulation. The enzyme behaves as if it is playing a geometrical game: compressing the reactive atoms into positions where a light nucleus can cross a barrier in a way that a heavier nucleus cannot.

To make that intuition visual, it helps to see two rate curves side by side. The plot below is schematic rather than a specific dataset, but it captures the essential experimental signature: hydrogen transfer remains fast across physiological temperatures while deuterium transfer is markedly slower, producing a large, diagnostic gap.

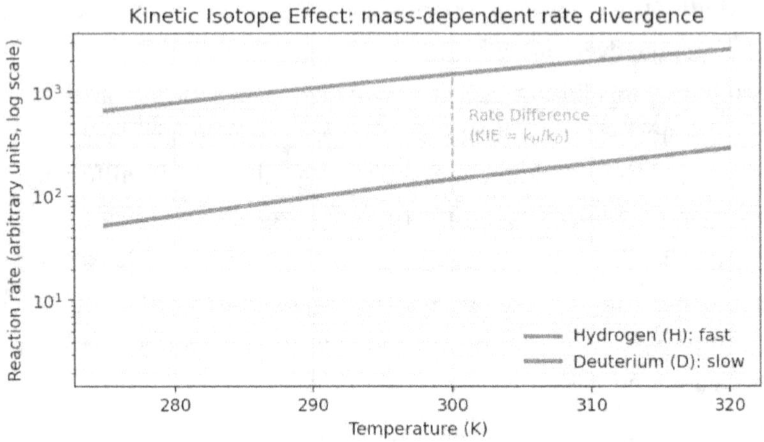

The virtue of isotope experiments is that they do not require one to see the proton tunneling directly. They exploit a basic feature of quantum mechanics: tunneling probability depends strongly on mass. Change only the mass while keeping the chemical identity nearly the same, and the quantum contribution changes while much of the classical chemistry stays put.

But enzymes are not only about the chemistry of a single bond. They are also about how the whole protein coordinates to make that chemistry possible. This is where the story becomes more subtle and, in a sense, more interesting. Evidence for tunneling can coexist with disagreements about what tunneling *means* for catalysis.

One influential arena for this debate is dihydrofolate reductase (DHFR), an enzyme essential for DNA synthesis because it helps generate the reduced folate needed for building nucleotide bases. DHFR catalysis includes a hydride transfer step. Experiments on *E. coli* DHFR measured hydrogen/deuterium isotope effects and found temperature-dependent behavior and Arrhenius parameter patterns that were often interpreted through tunneling models coupled to protein motion. Even more striking were the effects of remote mutations: change an amino acid far from the active site, one that does not alter the chemistry directly, and the catalytic efficiency can shift significantly. Such findings encouraged the idea that protein dynamics are not just background noise; they can gate access to the reactive configuration that makes hydrogen transfer efficient.

The word *gating* captures a key idea. The enzyme does not need to push a hydrogen over a barrier by brute energetic force. It needs to achieve, often fleetingly, a donor—acceptor arrangement in which the barrier is narrow enough for tunneling probability to rise. In that picture, protein motions are part of the catalytic machinery. They are not random flailing; they are statistically biased movements shaped by evolution so that the enzyme frequently visits the geometries where the chemical step can occur.

And then came an elegant complication: *heavy enzymes*.

In a heavy-enzyme experiment, one does not replace hydrogen with deuterium in the substrate. Instead, one builds the entire enzyme with heavier isotopes, typically by growing the organism (or expressing the protein) in media enriched in heavier carbon (^{13}C), nitrogen (^{15}N), and sometimes deuterium. The overall structure is essentially the same, but the vibrational frequencies of the protein shift because heavier atoms move differently. If fast protein vibrations are essential to promoting tunneling, then making the enzyme heavier might be expected to slow the reaction in a way that points to a dynamic coupling between protein motion and the chemical coordinate.

Some heavy-enzyme studies on DHFR did find slower rates for the heavier protein, supporting the idea that enzyme mass and therefore dynamics matter. But careful analysis and simulation suggested an important alternative interpretation: the observed slowing could be explained without invoking a large change in tunneling probability or barrier height. Instead, it could arise from changes in how trajectories cross and recross the transition-state region, from differences in environmental coupling, or from altered pre-exponential factors in the rate expression. In other words, the protein's mass can influence kinetics through several channels, not all of which mean "the enzyme is actively promoting tunneling."

This is where the debate in quantum enzymology often becomes semantic, and therefore surprisingly heated. There are at least three distinct questions that can be accidentally merged.

First: does hydrogen tunneling occur in enzyme-catalyzed reactions under physiological conditions? For many enzymes involving hydrogen transfer, the weight of isotope evidence makes it hard to deny that tunneling contributes at least to some degree.

Second: do enzymes *actively promote* tunneling beyond what would happen in a comparable reaction environment? Here the burden is heavier. It requires showing that the protein's dynamics and architecture are not merely permissive but specifically tuned to increase tunneling probability.

Third: is tunneling the main reason enzymes are so fast compared to uncatalyzed reactions in water? Many careful researchers answer no, or at least "not in general." They argue that most of the catalytic power comes from classical factors like electrostatic preorganization and transition-state stabilization, with tunneling modifying the chemical step in particular cases. Others argue that in certain hydrogen-transfer reactions, especially those with enormous isotope effects, tunneling is not merely a tweak but a central feature of how the enzyme achieves its rate.

A prominent dissenting viewpoint, associated with researchers who emphasize electrostatic and free-energy frameworks for catalysis, has pressed an important point: tunneling can be real without being *the source* of enzymatic rate enhancement. The enzyme can lower activation free energies in essentially classical ways, and tunneling can then occur as a natural consequence of hydrogen being light. From this perspective, talking about tunneling as the "secret" of enzymes risks confusing a mechanism of barrier crossing with the deeper reason the barrier is shaped the way it is in the first place.

This dissent is not an attempt to erase quantum behavior. It is a demand for careful accounting. If an enzyme accelerates a reaction by a factor of 10^{12}, and tunneling contributes a factor of, say, 10^{2} to the chemical step under certain conditions, then tunneling matters, but it cannot carry the entire explanatory burden. The enzyme is still doing enormous classical work in organizing charges, stabilizing transition states, excluding water, and steering reactants. Conversely, if tunneling changes the rate by a thousandfold in a step that is already rate-limiting for a pathway, it can be biologically decisive even if it is not responsible for the full trillionfold difference between enzyme and water.

Isotope effects remain valuable precisely because they cut through this philosophical fog. They answer a narrower, more empirical question: how strongly does mass influence the rate-determining hydrogen motion, and how does that influence change with temperature and with perturbations to the protein? Those are questions experiments can speak to.

There is also a quiet lesson in humility here. The strongest enzyme experiments do not treat the protein as a rigid sculpture, nor do they treat quantum mechanics as a supernatural force. They treat the enzyme as matter that fluctuates, and they treat the proton as matter that is unusually sensitive to those fluctuations. The signature of tunneling is not that physics has been suspended. It is that a small nucleus, placed in a carefully engineered molecular environment, behaves according to quantum rules in a way that measurably alters the pace of life.

A single isotope substitution does not feel dramatic in the hand. It does not change a molecule's color or smell. Yet it can expose what is otherwise invisible: that, in some of the most important steps in metabolism, nature is not merely waiting for thermal luck. It is exploiting a probability wave, and the weight of that wave can be measured.

Chapter 6

Navigating the Invisible

Every year, billions of birds traverse the globe with a precision that rivals modern GPS. We descend into the retina of the European robin to discover a biological compass that relies not on iron filings, but on the delicate quantum dance of entangled electrons. By exploring how weak magnetic fields influence chemical reactions, we uncover nature's potential mastery of quantum measurement in a warm, wet environment.

6.1 The Robin's Secret

On autumn nights in northern Europe, the sky can look empty. No birds in the hedgerows, no shadows crossing the fields. And yet the air is busy. A European robin, small enough to fit in a cupped hand, may be climbing into the darkness on quick wingbeats, joining a stream of migrants that will travel hundreds or thousands of kilometres. The robin does not see the destination. It does not have a map in the human sense. Still, it departs with a confidence that seems almost reckless: it points itself in a seasonally appropriate direction and keeps going.

For much of the twentieth century, the most unsettling detail was not that birds migrate—people have watched that for millennia—but that they can keep a bearing when familiar landmarks are gone.

Clouds erase the stars. The horizon disappears. The wind shoves and swivels. If you place a migratory bird under an overcast sky, in a place it has never been, it can still choose a direction that makes biological sense. Something steady and global is being used as reference.

The Earth provides one obvious candidate: its magnetic field. A compass needle makes it visible, in a way. But magnetism feels like the wrong sort of influence to guide a living creature. The geomagnetic field is weak, roughly the strength that makes a refrigerator magnet feel unimpressive in your hand. Worse, the field is not a simple arrow pointing north; it is a three-dimensional vector that dips downward into the ground in the northern hemisphere and upward in the southern. And to make the paradox sharper, the energy associated with such a weak field at the level of single molecules is tiny compared with the jostling thermal energy inside a warm animal. At body temperature, everything is being rattled, kicked, and buffeted by the random motion of atoms. How can such a small physical cue survive the noise of biology, let alone be turned into reliable behaviour?

For years, the easiest explanation sounded like the most plausible: birds might carry little pieces of magnetic material, like iron minerals, acting as microscopic bar magnets. Nature has magnetite—a magnetic iron oxide—and some organisms genuinely use it. If a bird had magnetite grains connected to a nerve ending, then the field might tug on them and provide directional information. It is a satisfying picture because it is mechanical. One can imagine a magnet pulling on a hair cell, the way a weight pulls on a spring scale. No mysterious chemistry required.

The trouble is that behaviour has a way of refusing neat stories.

A turning point came from an experimental device with a gentle name: the Emlen funnel. In the 1960s, the ornithologist Stephen Emlen wanted a way to measure migratory direction in a controlled setting. If a migratory bird is caged during migration season, it does not sit calmly and accept captivity. It becomes restless at night, hopping and fluttering in a particular preferred direction—a behaviour known as *Zugunruhe*, literally "migratory restlessness." The Emlen funnel turns that restlessness into data.

The apparatus is simple: a cone-shaped paper-lined funnel, often with an ink pad or carbon paper arrangement, and a wire mesh top to keep the bird from escaping. Put the bird inside at night and dim the lights. The bird will hop up the sloping walls again and again, trying to launch itself. Each hop leaves a scratch mark. In the morning, the paper tells a story: a cluster of marks thickest in the direction the bird was trying to go.

A robin in autumn does not scratch randomly. It shows a bias, as though its body contains an intention. Rotate the direction of magnetic north in the room using coils that produce a controlled magnetic field, and the scratches rotate with it. This is the sort of result that makes an experimentalist sit back and check their wiring twice. The bird is not reacting to some unnoticed visual cue; it is responding to an invisible vector in space, and it responds in a way you can steer by turning a dial.

That alone might still fit the magnetite picture. A bar magnet in the beak would turn when the field turns; perhaps the bird reads that rotation and chooses the corresponding direction. But as the experiments became more careful, birds began to reveal a stranger set of rules.

One rule is that many birds do not seem to care about polarity in the way a classroom compass does. A compass needle points north; reverse the magnetic field and the needle flips. In behavioural tests, migratory birds often behave as if they are using an *inclination* compass rather than a polarity compass. Inclination means they distinguish "poleward" from "equatorward" using the tilt of the field lines, not "north" from "south" as opposite polarities. Reverse the field direction but keep the inclination the same, and the bird may not flip its behaviour the way a needle would. This is not a minor technicality. It implies that the sensor is not simply a tiny permanent magnet that mechanically aligns, because a permanent magnet is exquisitely sensitive to polarity. A mechanism that naturally produces an inclination compass begins to look less like a rock and more like a process.

Another rule is that light matters.

It is easy to miss how odd that is. Human compasses do not need illumination. Neither do magnetite grains. Yet European robins show magnetic orientation only under particular lighting conditions. In many laboratory studies, robins can orient under dim blue or green light, but become disoriented under red light or in darkness. The details depend on the exact experimental setup and the species, but the theme is robust: the magnetic sense behaves like something entangled with vision.

This pushes the mystery from the beak toward the eye. A mechanical magnet in the head should not care whether photons are present. But a photochemical sensor would. If magnetoreception is triggered or modulated by light, then the obvious place to look is the retina, the tissue that evolved to do chemistry with photons.

The idea that magnetism might be *seen* rather than *felt* sounds like fantasy until you remember what vision already is. We speak as if the eye is a camera, collecting images. In reality, the retina is a chemical lab spread across the back of the eyeball. Photons strike molecules and trigger electronic changes; those changes set off cascades of reactions; those reactions modulate currents in neurons. The brain never receives light. It receives signals about chemical events caused by light. If some of those chemical events are slightly sensitive to magnetic fields, then the brain might receive a pattern of signals that correlates with the bird's orientation relative to the field.

At that point the question is no longer "Does magnetism affect behaviour?" but "Which tissue is doing the transduction?" And how do you show it?

One approach is blunt, and biologists tend to dislike it for good reasons: you interfere with a candidate pathway and see what breaks. If the beak is the sensor, then disrupting beak nerves should disrupt magnetic orientation. If the eye is the sensor, then blocking visual input should do it. But living animals are not simple machines. Birds can compensate. They have more than one cue. A bird might orient magnetically when the stars are hidden and switch to a star compass when the sky is clear. A bird might use olfaction or landmarks in some contexts. Negative results can be ambiguous: perhaps you disrupted the pathway but the bird used another. The most convincing experiments are those that use the bird's own rules against it, isolating one cue at a time.

The Emlen funnel makes that possible. It creates a stripped-down world: no landmarks, no wind, controlled lighting, controlled magnetic field. In this world, you can ask questions that are almost philosophical: if you remove everything the bird might be using, what remains? And if you keep everything constant but change only the spectrum of light, does the compass survive?

Robins answered in a way that felt like a clue left intentionally. Under wavelengths associated with the activation of certain retinal photopigments, orientation appears; under other wavelengths, it collapses. If you take away the stars and keep the magnetic field steady, the bird still aims. If you keep the stars away and change the light from blue to red, the bird loses its aim. The same animal, the same season, the same funnel, the same magnetic field—but a different colour of dim light—and an entire navigational sense switches off.

A story from early magnetoreception work illustrates how stubbornly birds insist on their own sensory logic. In some experiments, researchers found that a bird's preferred direction could be shifted simply by changing the magnetic field during a training period, as if the animal recalibrated. The bird did not behave like a compass needle; it behaved like a nervous system trying to interpret a cue. The magnetic field was not merely pulling; it was *informing*. This distinction matters because it explains why the search for a magnetic "organ" was never just anatomical. The real organ might be a distributed computation: chemistry in the retina, wiring through the brain, behavioural output.

There was also a practical reason the eye hypothesis gained traction: claims about magnetite-based sensors in the beak became controversial. Some anatomical studies reported iron-rich structures associated with nerve endings in the upper beak, suggesting a magnetite receptor. Later work challenged those interpretations, arguing that some of the iron-containing cells were macrophages—immune cells that happen to store iron—not specialized magnetoreceptors. In a field that depends on tiny effects and delicate behavioural assays, a disputed anatomical claim can derail years of consensus. Meanwhile the light-dependence of the compass was stubbornly reproducible. The retina kept reappearing as the site with the most consistent behavioural fingerprints.

Still, light-dependence alone is not enough. The eye is full of photochemistry. What sort of chemical reaction would be sensitive to a magnetic field as weak as the Earth's? And why would that sensitivity depend on the direction of the field, not merely its strength?

Answering that requires a shift in imagination. The magnetic field is too weak to "push" chemical reactions over an energy barrier in the usual way. Yet chemical reactions are not only about energy landscapes. They are also about *paths* through fleeting intermediate states. Some of those intermediate states carry unpaired electrons. And unpaired electrons have a quantum property called *spin*. Spin is not literally a tiny ball rotating, but it behaves as if it provides a magnetic moment, like an intrinsic compass needle attached to the electron itself. When molecules briefly form pairs of radicals—molecules with unpaired electrons—the spins of those electrons can be linked in a correlated way, and their relative alignment can influence which chemical products are formed.

At the behavioural level, one does not need to know the full quantum story to appreciate the experimental logic. If a magnetic sense is light-dependent and likely retinal, then it should be possible to localize it further by asking whether one eye can do the job alone.

Bird eyes are not mere cameras on stalks. Each eye feeds information into a brain with lateralized processing. In many birds, the right eye has privileged access to certain visual pathways, and experiments have reported lateralization in magnetic tasks as well: sometimes the right eye seems more important for magnetic compass orientation. The details vary across studies and species, and lateralization is rarely absolute, but it offers a testable handle. Cover one eye with a small cap. Keep the lighting and magnetic field controlled. Does the bird still orient?

When researchers did this kind of experiment, what emerged was not a simple "yes" or "no" but a pattern consistent with a retinal sensor: orientation often persisted when one eye was available and degraded when visual input was compromised in the relevant spectral range. Birds are, after all, using their eyes constantly in captivity; covering an eye does not merely remove a magnetic sensor, it changes the whole visual world. But in carefully designed conditions, the dependence on light and the dependence on eye input converge on the same conclusion: the magnetic compass is deeply tied to vision.

For a general reader, it may help to picture what a "visual" magnetic sense could mean. No one thinks robins see glowing magnetic lines like a physics diagram. The plausible claim is subtler: magnetic information might appear as a faint modulation, an overlay on the normal visual scene. Imagine wearing sunglasses with a barely visible smudge that shifts position when you turn. You would not see magnetism itself, but you could use the smudge as a directional cue. In a bird, the "smudge" could be a pattern in retinal activity whose geometry depends on the angle between the bird's head and the magnetic field. The bird would learn, through evolution and experience, to interpret that pattern as "head this way."

The elegance of the hypothesis is that it matches several behavioural constraints at once. A retinal chemical mechanism can be light-dependent because photons initiate the chemistry. It can behave as an inclination compass because the underlying physics is naturally symmetric with respect to field reversal under many conditions. It can be disrupted by certain wavelengths because different wavelengths excite different photochemical pathways. And because the retina is spatially organized, it can convert a directional signal into a pattern.

All of this might have remained speculative if not for a particular kind of experiment that feels almost mischievous: rather than changing the static magnetic field, you disturb it with a tiny oscillating field at radio frequencies. If the bird's compass were a chunk of magnetite pulling on tissue, a weak radio-frequency field should do very little. A mechanical magnet has inertia; it does not care about a rapidly oscillating microtesla perturbation. But if the compass relies on the dynamics of electron spins in short-lived radical pairs, then radio-frequency fields tuned to the relevant spin transitions could scramble the signal.

When such radio-frequency "jamming" experiments were performed in the early 2000s by Thorsten Ritz and collaborators, the behavioural effect was striking: under conditions where birds oriented reliably, adding a weak oscillating magnetic field at particular frequencies could disorient them. The point is not that the experiment proves every molecular detail. The point is that it uses the physics of spin as a diagnostic. If you can jam the compass by poking it at the frequencies spins respond to, you have narrowed the list of plausible mechanisms dramatically.

The robin, in other words, is not merely responding to magnetism; it is responding to magnetism in a way that looks like quantum spin chemistry embedded in a sensory organ.

There is one more twist that makes the robin's secret feel uncomfortably close to human life. If the compass is that delicate, then it may be vulnerable to the electromagnetic clutter of modern environments. A city is filled with weak broadband electromagnetic noise from power lines, electronics, and communication devices. To us, this is invisible and mostly harmless at everyday levels, an engineering background hum. To a spin-sensitive chemical compass, it may be like trying to hear a whisper in a crowded room.

Experiments in the 2010s associated with Henrik Mouritsen's group reported that European robins could become disoriented in the electromagnetic conditions typical of urban settings, and that shielding the birds from that noise could restore orientation. The ecological implications are still discussed cautiously—migration is a robust behaviour shaped by many cues, and laboratory conditions are not the open sky—but the result adds a sobering realism. If a robin's magnetic sense is truly tied to subtle spin dynamics, then the modern world may be changing not only landscapes and climates, but also the sensory texture of the air.

None of this makes the phenomenon less wondrous. It makes it more specific.

A robin in an Emlen funnel is not performing magic; it is revealing an interface between physics and biology that evolution has refined. The bird scratches at paper because it wants to leave. The scratches accumulate into a direction because its nervous system has found a reliable reference when other references vanish. And the reference is not a crude internal bar magnet. It is a light-dependent process—one that lives where photons are already being turned into perception.

The simplest way to hold the paradox in mind is to imagine what the bird is actually doing on a migratory night. It is not measuring magnetism the way a laboratory instrument measures magnetism. It is running a living algorithm: sampling a weak, global cue; combining it with time of year and internal state; translating it into a behavioural decision; and then repeating, for hours, through cold air and changing winds. The Earth offers a field so faint we rarely think about it. The robin treats that faintness not as a barrier but as a feature, extracting direction from something most animals ignore, as though the planet itself is quietly providing a scaffold for movement.

6.2 The Chemical Compass

A compass made of chemistry sounds like a metaphor until it starts behaving like an instrument. The robin's magnetic sense depends on light, and that single fact forces the imagination into an unfamiliar shape. Whatever the bird is reading is not a tug on a grain of iron but a photochemical event, something that begins when a photon is absorbed and ends as a nervous signal. The startling part is that magnetism, which seems like a slow, sturdy, iron-and-needles kind of phenomenon, can reach into chemistry at all.

Chemists have long known that magnetic fields can influence reactions, but usually in ways that require fields far stronger than the Earth's. A laboratory magnet that can lift a wrench is an entirely different beast from the geomagnetic field that quietly threads through your body right now. So the question becomes precise: what sort of photochemistry could be sensitive to a field of only a few tens of microtesla, and sensitive not just to *strength* but to *direction*?

The answer that has gradually taken the lead is called the *radical pair mechanism*. The name is accurate and a little misleading. It is accurate because it really does involve radicals, molecules or molecular fragments with an unpaired electron. It is misleading because "radical" suggests something violent or unstable. In biology, radicals are often treated as troublemakers, associated with oxidative damage. Yet radicals are also ordinary intermediates in controlled reactions. Cells manage them the way a kitchen manages heat: dangerous if uncontrolled, indispensable if handled well.

The radical pair mechanism begins with a moment so small it seems beneath the dignity of animal behaviour: one electron jumps.

Consider a protein that can absorb blue light. It holds a light-absorbing cofactor—a helper molecule—that acts as a springboard for electron transfer. When a photon is absorbed, the cofactor is lifted into an excited state. In that excited state, an electron can move from one part of the system to another, leaving behind two radicals: one where the electron departed, one where it arrived. These two radicals are created together, in a single event, and for a brief time they remain a linked pair.

That linkage is the essential trick. The two unpaired electrons do not merely exist; they exist in a correlated *spin state*. Spin is a quantum property, often introduced with the metaphor of a tiny spinning top, but what matters here is simpler: an unpaired electron behaves like a tiny magnet with an orientation, and quantum mechanics allows two such "tiny magnets" to be correlated in specific ways.

There are two basic combined states to care about. In the *singlet* state, the two spins are anti-aligned in a way that gives the pair a total spin of zero. In the *triplet* state, the spins are aligned (or effectively aligned in one of three related configurations) so the pair has a total spin of one. These are not just labels. Chemical reactions can care deeply about them because certain reaction pathways are allowed from one spin state and forbidden from the other. A radical pair in a singlet state may recombine easily, snapping back into a stable molecule. The same radicals in a triplet state may be blocked from recombination and instead proceed to form different products.

At first glance, this might sound like an on/off switch that could power a sensor. Create a radical pair with light. Let magnetism influence whether it is singlet or triplet. Read out different chemical products. Direction becomes chemistry. Chemistry becomes perception.

The hard part is hidden in the word "influence." The Earth's field is weak. It cannot simply *force* a pair into singlet or triplet. Instead it biases a natural internal motion called *singlet–triplet interconversion*. Once a radical pair is created, its spin state does not necessarily remain fixed. The spins precess, wobble, and mix under the combined influence of the external magnetic field and internal magnetic interactions within the molecule. Those internal interactions mainly come from nearby atomic nuclei, whose own spins create tiny magnetic fields that the electron spins feel. This is called *hyperfine coupling*. "Hyperfine" is an intimidating word for a modest idea: the electron's magnetic character interacts with the magnetic character of neighbouring nuclei, and that interaction depends on geometry and local chemical structure.

Here is the key point: hyperfine coupling can be *anisotropic*—direction-dependent. In other words, the way the electron feels the nuclear environment can change when the molecule is rotated relative to the external magnetic field. The radical pair's spin dynamics are then subtly different depending on orientation. Over the lifetime of the radical pair, those differences translate into different probabilities of ending in the singlet channel or the triplet channel. If the radical pair lives long enough for this mixing to matter, the chemical yield becomes a function of angle relative to the field.

This is why the mechanism is often described as a *chemical compass*. It does not measure magnetic north like a needle. It converts the direction of the magnetic field into a difference in chemical outcomes.

A historical anecdote makes the intellectual leap clearer. Around 2000, Thorsten Ritz and collaborators put forward a detailed model of how a photoreceptor-based magnetic compass could work, proposing not only that a radical pair could be sensitive at Earth-field strengths but also that its output could be linked to visual processing. The audacity was not in invoking quantum mechanics—chemists use it every day—but in claiming that such delicate spin chemistry could be embedded in a biological sensor and survive long enough, reliably enough, to guide an animal at night. The proposal helped organize a field that had been pulled between two intuitions: magnets in tissue versus chemistry in the eye.

The most plausible biological host for this photochemistry is a family of proteins called *cryptochromes*. The name comes from their original discovery as "cryptic" blue-light receptors in plants, but cryptochromes are widespread: they appear in plants, insects, and vertebrates, and they are involved in circadian rhythms as well as light sensing. A cryptochrome is a protein with a built-in light-harvesting cofactor, typically flavin adenine dinucleotide (FAD), a molecule familiar from cellular metabolism. FAD can absorb blue light and participate in electron transfer reactions.

In the cryptochrome picture of magnetoreception, blue light excites the flavin cofactor. An electron is transferred along a chain of amino acids (often involving tryptophan residues), creating a radical pair involving the flavin and its partner radical. Because the radicals are created together, their spins begin in a well-defined correlated state. From there, singlet–triplet interconversion begins, influenced by hyperfine interactions and the external magnetic field. Different spin histories lead to different chemical fates: recombination, longer-lived signaling states, or other downstream chemical differences. The bird does not need to "understand" any of this. It only needs a reliable link between magnetic direction and some pattern of neural activity.

The insistence on blue light now makes a kind of biochemical sense. If cryptochrome activation is required, then lighting conditions that fail to excite the flavin effectively will also fail to start the radical pair dynamics. Red light may simply be the wrong energy to trigger the relevant photochemistry. Darkness means no photons, no excited flavin, no radical pairs, no compass signal.

But a second requirement is less obvious: *ordering*. If the chemical compass depends on anisotropy, and anisotropy depends on molecular orientation, then the molecules cannot be tumbling completely randomly. In a beaker, molecules rotate freely; any direction-dependent effect averages away. In a protein embedded in a membrane, or organized in a structured tissue, molecules may be partially aligned. The retina is not a stirred solution. It is a layered, organized sheet in which proteins can have preferred orientations relative to the geometry of the eye. That organization offers a way for direction-dependent spin chemistry to survive averaging.

This is also why the proposed sensory experience is often described as a pattern. If cryptochromes across the retina are oriented in a systematic way, then the yield of a signaling state could vary across the retinal surface depending on the angle between the head and the magnetic field. The outcome could be a weak but structured modulation of retinal activity, something like a compass rose painted in chemicals rather than ink.

At this point, a skeptic usually asks the most reasonable question available: even if the chemistry is direction-dependent, can it be strong enough to matter? How do you turn a small change in product yield into a robust signal that a nervous system can read?

One answer is that biology is good at amplification, not in the crude sense of making a small voltage larger, but in the kinetic sense of turning a small bias into a large difference over time. The radical pair event happens fast, but it can feed into reaction networks that include slow steps, feedback, and accumulation. Under continuous illumination, a small difference in the fraction of molecules entering one pathway versus another can lead to a measurable difference in the concentration of a signaling state. The system does not need a huge magnetic effect per photon; it needs a consistent bias that can be integrated.

A useful analogy is not an electronic amplifier but a voting process. If each photon "casts a vote" for singlet or triplet outcomes, the magnetic field does not need to force unanimous decisions. It needs only to tilt the odds slightly. Over many photons and many molecules, that slight tilt can become a reliable statistical signal, especially if downstream chemistry stabilizes one outcome or prolongs its lifetime.

Cryptochromes offer plausible molecular features for such integration. Some cryptochrome photocycles involve relatively long-lived radical or semiquinone states. If one spin pathway leads to a state that persists longer, it can disproportionately influence subsequent biochemical interactions, even if the initial yield difference is modest. Modeling work has shown that networks of reactions can amplify magnetic field effects by coupling fast spin-dependent steps to slower chemical bottlenecks, effectively letting time do the heavy lifting.

Evidence has also been shifting from plausibility to direct measurement. In 2021, work led by Jingjing Xu and Lauren Jarocha reported magnetic sensitivity in vitro for cryptochrome 4 from the European robin, with photochemistry consistent with radical-pair spin dynamics. In vitro is not in vivo, and a protein in solution is not a retina in a living bird. But it matters enormously to show that a candidate protein can, under controlled conditions, exhibit magnetic-field-dependent behaviour in the right direction and at the right field strengths. It transforms a mechanism from a clever story into a measurable physical effect anchored in a real biological molecule.

Still, there is a lingering conceptual stumbling block that deserves to be handled honestly. The energy scale of the geomagnetic field is tiny. In a warm cell, thermal energy is enormous compared with the Zeeman splitting associated with microtesla fields. How can such a weak field bias anything at all?

The radical pair mechanism sidesteps the energy argument because it does not rely on a magnetic field making one state energetically favourable in a Boltzmann sense. It relies on coherent dynamics in spin space and on *selection rules* that govern which spin states can react along which pathways. The magnetic field changes the *time evolution* of the spin state—how singlet character converts to triplet character and back—and reaction rates then "sample" that evolving mixture. In other words, the field is not paying an energetic price to push a chemical boulder uphill; it is altering the choreography of a dance that was happening anyway, so that when the music stops the dancers are more likely to be in one arrangement than another.

Because that choreography depends on orientation, and because the bird can rotate its head relative to the field, a directional signal becomes available. The outcome is also naturally symmetric in a way that fits an inclination compass: many radical pair implementations are sensitive to the axis of the field rather than its polarity, because reversing the field does not necessarily change the relevant mixing dynamics in a way that produces a distinct signal. A molecule is not a compass needle; it is a system of coupled spins and reactions.

To keep the mechanism grounded, it helps to walk through a single photochemical "event" in a robin's retina as if it were a tiny drama.

A photon of blue light enters the eye and is absorbed by the flavin cofactor in cryptochrome. The flavin is promoted into an excited electronic state. Almost immediately, an electron is transferred, creating two radicals. The radicals are born together, their spins correlated like dancers starting in a synchronized pose. Internal magnetic interactions—hyperfine couplings to nearby nuclei—begin to twist that pose, mixing singlet and triplet character. The Earth's magnetic field, weak but steady, nudges the rate and manner of that twisting, in a way that depends on the cryptochrome's orientation within the retinal tissue. Within microseconds (or less, depending on the details), chemistry "reads out" the spin state: singlet-like pairs may recombine; triplet-like pairs may separate and lead to a different chemical intermediate. That intermediate might be longer-lived, might interact with other proteins, might influence ion channels or signaling pathways. Across millions of cryptochrome molecules, those small biases sum into a faint but structured pattern of biochemical activity. The bird's nervous system, trained by evolution to treat that pattern as information, adjusts its behaviour: a preference in scratch marks on paper, or a steady heading under the stars.

It is hard not to feel a kind of vertigo at the thought that the Earth's field is being turned into biochemistry through quantum rules. But it is also strangely fitting. Life has always been opportunistic. It uses what works. If electron spins offer a way to turn a global physical reference into a sensory cue, then evolution has had millions of years to refine the details: which cryptochrome variant, which electron-transfer chain, which tissue arrangement, which downstream amplification.

A final detail returns the story to the robin itself. If the compass is a chemical process driven by light, then it is not a rigid instrument but a living one. It can adapt, recalibrate, and perhaps even be fooled. Changes in ambient magnetic intensity can shift the "functional window" in which birds orient best. Lighting conditions can modulate performance. And the chemical compass can be vulnerable to interference from oscillating fields, not because it is fragile in a hand-waving sense, but because its working parts are spins, and spins respond to magnetic perturbations in ways that classical needles do not.

There is a peculiar intimacy in that. The robin does not carry the Earth's magnetic field like a tool. The field passes through the bird the way it passes through everything else. The bird's achievement is to have built, inside a thin sheet of living tissue, a molecular system that notices the field's direction and converts it into a usable hint. Blue light, a protein, an electron hop, and the quiet insistence of a planet-sized magnet together create a sense that feels almost like an extra dimension of vision.

6.3 Spins in the Noise

A robin's compass, if it truly runs on radical pairs, is built out of the most skittish parts of matter. An electron's spin is a superbly sensitive magnetic pointer, but it is also a gossip: it "hears" every nearby magnetic influence, including the random ones produced by vibrating atoms and tumbling molecules. Biology is not a vacuum flask. It is warm, wet, crowded, and constantly in motion. That is why the phrase *quantum biology* can provoke an immediate wince in a physicist. Quantum effects are famous for being delicate. Living tissue is famous for being noisy. The marriage looks doomed.

The paradox is sharpened by a number that is almost comical in its mismatch. The energy associated with aligning a single electron's magnetic moment with the Earth's field is tiny. Thermal energy at room temperature is often summarized as $k_B T$, and it is large enough to smear out many small energetic preferences. If magnetoreception were trying to work by straightforward energetics—"north-facing is lower energy than south-facing"—it would fail before it started. Any preference would be drowned in the ceaseless jitter of thermal motion.

The radical pair mechanism avoids that trap by refusing to play the energy game. It uses dynamics and rules: how spins evolve in time, and which chemical steps are allowed from which spin states. Still, dynamics can be destroyed by noise. If the spins lose their delicate correlation too quickly, the compass signal becomes a blur. This is where the idea of *coherence* enters.

Coherence is an overloaded word, sometimes used as a synonym for any kind of quantum weirdness. Here it means something quite concrete: the spin state of a radical pair retains a well-defined phase relationship long enough for singlet and triplet character to interconvert in a way that depends on the external magnetic field. Without that, the system becomes classical in the dullest sense. It still has spins, but they behave like coin flips that forget their past instantly. In that case, the chemical outcomes become insensitive to the details of spin evolution, and the direction information disappears.

A helpful image is to imagine two tiny dancers (the two electron spins) trying to perform a coordinated routine while surrounded by a crowd that keeps bumping into them. If the bumps are gentle and infrequent, the dancers can keep time with each other. If the bumps are constant, the choreography collapses into random shoving. The radical pair compass needs the choreography, because the external magnetic field influences the timing and style of the dance.

What, exactly, is doing the bumping in a cell? Many things. The most obvious are collisions and vibrations: molecules in solution rotate and translate, proteins flex, water molecules flicker into new orientations. Then there are magnetic bumps: nearby nuclei have spins too, and their magnetic moments create fluctuating local fields. Even the electron spins of other radicals can contribute. In physical terms, these sources of noise cause *spin relaxation* and *decoherence*. Relaxation is the tendency for spins to lose a preferred orientation. Decoherence is the loss of phase relationships that allow genuinely quantum interference between different spin evolutions. Both processes smear out the subtle differences the compass tries to detect.

At first glance, this looks fatal. The radical pair needs to live long enough—often envisioned as microseconds—for the geomagnetic field to matter. That is an eternity on molecular timescales, and an eternity inside a warm protein would seem implausible. This objection has been voiced repeatedly, and it is healthy. The field has benefited from it, because it forces every claim to confront the world as it is, not as a theorist would like it to be.

Yet the story is not simply "coherence good, decoherence bad." Realistic modeling has shown that some forms of relaxation can, in certain parameter regimes, *improve* compass performance. That sounds like heresy if one's mental model is a pristine quantum computer that fails as soon as it meets the slightest noise. But a biological sensor is not trying to preserve an arbitrary quantum state indefinitely. It is trying to convert a physical parameter into a reliable chemical bias.

Noise can, in some cases, help by preventing the system from getting stuck in unhelpful states, by broadening sensitivity, or by effectively filtering out irrelevant fluctuations. The point is not that noise is desirable in general; it is that the relevant question is functional: what mixture of coherence and relaxation produces the strongest, most robust directional signal under biological conditions?

This functional view also changes what counts as a "smoking gun." If one cannot easily measure coherence in a living bird's retina, one can instead look for behavioural signatures that are hard to explain without spin dynamics.

The most famous of these signatures comes from a kind of experiment that feels like an act of sabotage. If the compass depends on electron spins, then it should be vulnerable to oscillating magnetic fields at frequencies that resonate with spin precession. If the compass depends on magnetite grains tugging on nerve endings, the same oscillating fields should have little effect, especially if they are extremely weak.

The relevant physics is familiar to anyone who has seen a child push a swing. Push at the right rhythm and a small force builds a large motion. Push at the wrong rhythm and nothing much happens. Electron spins, placed in a magnetic field, precess at a characteristic frequency proportional to field strength. This is the basis of electron paramagnetic resonance (EPR) spectroscopy, a standard tool in chemistry and physics. In a laboratory, one uses microwave or radio-frequency fields to perturb spins and read out their behaviour.

In the early 2000s, Thorsten Ritz and collaborators brought this logic into the biology lab. They tested migratory birds under controlled magnetic fields and then applied weak radio-frequency magnetic fields designed to interfere with the expected spin dynamics of radical pairs. The result, reported in 2004, was arresting: under conditions where birds oriented reliably, certain oscillating fields could disrupt their orientation. The animals became confused in a way that looked less like a general stress response and more like a broken compass. It was as if the compass had been jammed.

The force of this result is not in its complexity but in its specificity. Weak radio-frequency fields at the appropriate frequencies are expected to perturb spin dynamics. They are not expected to mechanically twist magnetite grains in any efficient way. A magnetite particle can respond to a static field, but a rapidly oscillating microtesla field is like a whisper trying to shove a boulder back and forth thousands of times per second. The behavioural disruption therefore points strongly toward a mechanism in which spins—not rocks—are the sensitive element.

Even more provocatively, later work suggested that birds can be disrupted not only by carefully tuned laboratory signals but also by broad-band electromagnetic noise of the sort that modern environments produce. Experiments associated with Henrik Mouritsen's group (2014) reported that European robins tested in conditions resembling urban electromagnetic background could lose their magnetic orientation, and that shielding could restore it. Whether and how this plays out in the complexity of real migration remains an active area of debate, but the laboratory message is consistent: the underlying sensor is exquisitely sensitive to weak oscillatory magnetic fields. That level of sensitivity is very hard to reconcile with a purely mechanical magnetite compass, and uncomfortably easy to reconcile with spin chemistry.

This is where coherence returns, because susceptibility to oscillating fields provides constraints on the timescales involved. A system that forgets its spin correlations almost immediately would not be strongly perturbed by a weak oscillating field; it would be too noisy already. Conversely, a system that retains coherence for long enough can be thrown off rhythm by the right kind of external perturbation. Behaviour becomes a window into microscopic time.

There is a temptation to turn this into a simple number: "the bird must have x microseconds of coherence." Reality is messier. Radical pair dynamics involve multiple interactions: hyperfine couplings, electron-electron interactions, g-factor anisotropies, and relaxation processes. The radical pair may not be a single pair but a sequence of pairs along an electron-transfer chain, as suggested in cryptochrome photochemistry. The readout might not depend on perfect coherence but on the interplay between coherent evolution and stochastic events like recombination or spin-selective reactions. So the behavioural constraints do not translate into one neat coherence time. They translate into a region of plausibility in a space of parameters, and that region is increasingly being explored with a blend of modelling, spectroscopy, and behavioural work.

A concrete way to keep this grounded is to think about what the bird is actually doing with the signal. The retina, if it hosts the relevant chemistry, is flooded with photons during the day but is operating at low light levels at night. The compass has to work in dim, blue-leaning conditions where cryptochrome photochemistry can be initiated but the overall signal may be small. The nervous system must integrate information across space (the retinal surface) and time (many photochemical events), while the bird moves its head and body. Any useful magnetic pattern must be stable enough across these movements to guide behaviour, but flexible enough to recalibrate when conditions change.

This functional pressure may explain why the mechanism might not require what a physicist would call "long-lived quantum coherence" in the heroic sense. The radical pair does not need to preserve a superposition forever. It needs just enough quantum correlation to let the Earth's field bias singlet-triplet interconversion before chemistry commits to an outcome. If downstream kinetics can amplify small biases, the sensor can tolerate more noise than one might initially fear. Chemical amplification, in this context, is not an afterthought; it is the bridge between a faint spin effect and an animal's decision to fly southwest rather than southeast.

There is also a second layer of noise that is easy to overlook: biological *variability*. No two proteins are identical in their microenvironment. Cryptochromes in different cells may be oriented slightly differently, embedded in membranes with varying composition, surrounded by different local fields from nearby molecules. Even within one cell, proteins flex and breathe. A compass based on spin chemistry must therefore work not only in the face of thermal agitation but also in the face of manufacturing tolerances. Evolution has solved similar problems elsewhere by using redundancy and averaging: many receptors contributing to one percept, many weak signals summing into a robust one.

The radical pair mechanism fits naturally into this biological style. It is inherently statistical. Each photochemical event is a tiny trial. The magnetic field biases the probabilities. The retina provides many trials in parallel. The nervous system reads out a bias, not a deterministic outcome.

At the same time, the RF disruption experiments remind us that the system cannot be *too* sloppy. If the relevant spin dynamics were washed out almost instantly, there would be nothing coherent to disrupt. The behavioural "jamming" suggests that the compass sits in an intermediate regime: coherent enough to be steerable and disruptable, noisy enough to be biologically plausible.

This intermediate regime also offers a deeper lesson about the boundary between quantum and classical. The usual story is that quantum effects belong to the microscopic world and classical behaviour emerges when systems get large and warm. Magnetoreception, if the radical pair mechanism is correct, shows a different possibility: quantum effects can survive in a limited, task-specific way inside a warm organism, not as a general-purpose quantum computer but as a specialized sensor that exploits just one quantum degree of freedom—spin—and then hands off the result to chemistry and physiology as quickly as possible.

A small but telling detail is that the external field does not need to be strong if the system is designed to respond to it. An EPR spectroscopist can detect tiny magnetic influences because the experiment is built around spin sensitivity. Biology, in this view, has built an EPR-like sensitivity without the laboratory apparatus, using proteins as scaffolds and reaction kinetics as readout electronics. The robin is not doing spectroscopy consciously. It is carrying out a piece of physics because it has proved useful.

There remains room for pluralism. Evidence increasingly suggests that birds may use more than one magnetic sense. A light-dependent inclination compass appears tied to the eye and consistent with radical pairs. Separately, there are indications of magnetite-based pathways associated with the trigeminal system that may contribute to sensing field intensity for "map" information, the kind that might help an animal recognize where it is rather than which way to go. Multiple mechanisms are not a weakness. They are what evolution tends to produce when the problem is important and the environment is variable. Human navigation is not one sense; it is vision, vestibular cues, proprioception, memory, and inference stitched together. A migrating bird's navigation is likely similar in spirit.

Yet the resonance disruption results remain unusually hard to ignore. They are not simply correlations. They are interventions guided by specific physics. That is why they are often treated as the strongest behavioural evidence that magnetoreception involves quantum spin dynamics. A field of biology that could easily have been trapped in vague language about "mysterious forces" instead found itself being tested with the logic of resonance.

One can feel the strangeness of this without exaggeration. In a lab at night, a small bird hops in a paper funnel, driven by an ancient urge to migrate. The experimenters apply a whisper of radio-frequency magnetism—far too small to move a needle compass across the room—and the bird's internal certainty falters. Not because the bird is weak-willed, and not because the field is strong, but because something inside its eye is trying to keep time with spins, and the rhythm has been disturbed.

The most unsettling possibility is not that birds use quantum mechanics. The unsettling possibility is how ordinary that might be. Spins are ubiquitous. Radical pairs appear in many photochemical reactions. Cryptochromes are widespread. The ingredients are not exotic; the organization is. The robin's achievement is to hold a fleeting quantum sensitivity long enough to matter, then to bury it immediately under layers of chemistry and neural processing so that an animal can use it without ever noticing the physics. The night sky remains dark, the hedgerows remain silent, and the bird keeps a heading by listening, in effect, to the quiet precession of electrons against the planet's field.

Chapter 7

The Quantum Eye

Beneath the retina of a migrating robin lies a mechanism more sensitive than most human engineering, relying on the delicate interplay of subatomic particles. This chapter descends into the molecular machinery of the avian eye to discover how fleeting pairs of entangled electrons might translate the Earth's weak magnetic field into a perceptible visual overlay.

7.1 The Cryptochrome Candidate

A migratory bird does not carry a compass in its beak. Nothing in its skull clicks north like a toy needle. And yet, season after season, a robin leaving northern Europe can set a night-time course toward the Mediterranean with a reliability that makes human navigation look, by comparison, like an improvisation. The puzzle is not only *what* information the bird uses, but *where* in the body that information first becomes a usable signal.

The earliest guesses were delightfully literal. If the Earth has a magnetic field, then perhaps an animal must contain a magnetic material: tiny iron-rich crystals acting like microscopic bar magnets. Magnetite, an iron oxide that can hold a permanent magnetic moment, exists in many organisms and is undeniably real biology. It promised a straightforward sensor: a grain that tries to twist in the geomagnetic field, tugging on a membrane, opening a channel, firing a neuron. The appeal is obvious: a classical compass made of biological parts.

But the behavior of birds kept hinting that any such picture was incomplete. The magnetic sense seemed tied to light and to the eyes. Birds oriented differently in different colors of illumination, sometimes losing their magnetic ability under red light while keeping it under blue-green light. If that clue is taken seriously, it pushes the search away from iron crystals in the beak and toward something that lives where photons are routinely absorbed: the retina.

That shift sounds simple, but it is a radical move in biological thinking. The retina is usually treated as a camera: capture light, translate it into electrical signals, build a picture of the world. Now imagine it as a chemical laboratory running a second, parallel experiment in the same tissue: using light not to form an image but to create a fleeting molecular state whose outcome depends, subtly, on the direction of the Earth's magnetic field. The eye becomes not merely an optical instrument but a physical one.

The name most often attached to this idea is *cryptochrome*. Even the word has the feel of a clue left in plain sight: *crypto* for hidden, *chrome* for color. Cryptochromes are proteins found across much of life. In plants they help regulate growth and development in response to blue light. In animals they are famous for their connection to circadian rhythms, the internal daily clock. They belong to a broader family of light-sensitive proteins (flavoproteins) that carry a molecular cofactor called flavin adenine dinucleotide, usually abbreviated FAD. FAD is not exotic; it is a workhorse of metabolism, a small molecule that can accept and donate electrons. The cryptochrome protein is, in effect, a carefully built cradle for this cofactor, positioning it amid a scaffold of amino acids so that light can trigger controlled electron transfers.

The great attraction of cryptochrome, for anyone looking for a magnetic compass mechanism, is that it naturally performs the right kind of chemistry. The molecular compass hypothesis does not ask for a bar magnet. It asks for something more delicate: a reaction whose products depend on the quantum spin states of a pair of electrons. To even entertain that idea, a protein must do two things. First, it must generate a pair of unpaired electrons at the right moment, because electron spins are only interesting in this context when electrons are unpaired, living in radicals. Second, it must do so in a repeatable, structured way inside a warm, wet cell, where random collisions and thermal noise usually erase delicate quantum features almost as soon as they appear.

Cryptochrome offers a plausible route to both. When a cryptochrome absorbs a photon in the blue part of the spectrum, the energy is not used to heat the protein up in some vague way; it is used to rearrange electrons. In the simplest sketch, the photon excites the flavin cofactor. That excitation makes it energetically favorable for an electron to hop along a chain of amino acids in the protein, often involving several tryptophan residues (tryptophan is one of the standard amino acids, and it is unusually good at participating in electron transfer). The result is a pair of radicals: the flavin becomes a radical, and one of the tryptophans becomes a radical, separated by a distance that is small on human scales but enormous to an electron. Two radicals, two unpaired electrons, created in a defined geometry and on a timescale set by the protein's internal wiring.

That sounds like a tidy, mechanistic picture, but it lands immediately on an apparently devastating objection. Quantum states are fragile; everyone knows this in the modern, everyday sense. A phone call drops when the signal is noisy. A violin's note is spoiled by a careless touch. In the quantum world, "noise" is not merely a nuisance but the default condition. The interior of a cell is a crowded bath of water molecules, ions, vibrating bonds, and jostling proteins. If a quantum effect is going to matter, it must not only exist, it must be *engineered* by evolution to survive long enough to influence chemistry in a consistent way.

The key is to be clear about what kind of quantum state is being invoked. The candidate mechanism does not require a macroscopic superposition, a bird being in two places at once. It requires something much smaller and more ordinary in quantum chemistry: coherent spin dynamics within a radical pair. When two radicals are created together by an electron transfer, the spins of their unpaired electrons are not independent coins to be flipped at random. They begin their joint life correlated, often in a particular combined spin state. The language used to describe those correlations can become slippery, but the underlying point is concrete: the joint spin state influences which chemical pathways are allowed.

Chemistry students learn selection rules early on, usually in a form that seems like technical bookkeeping. Some reactions proceed efficiently from a singlet state (two electron spins paired anti-parallel), others from a triplet state (spins aligned in one of three parallel-like configurations). In many radical-pair reactions, interconversion between singlet and triplet character controls whether the radicals recombine to their original molecule or drift toward a different product. This is not speculative; it is a well-established part of spin chemistry in test tubes. The daring proposal is to place this known physics inside a protein in a retina and to let the Earth's magnetic field, weak as it is, bias the interconversion just enough that a living system can read out a direction.

In that picture, cryptochrome has to act like a stage manager for a microscopic drama. The protein holds the radicals at a particular separation. It sets the local magnetic environment experienced by each electron through nearby atomic nuclei, whose own spins create tiny internal magnetic fields (the hyperfine interactions familiar from magnetic resonance). It limits which motions are available during the crucial microseconds, so the radical pair is not immediately randomized by collisions. It may even anchor the radical pair in a fixed orientation relative to the structure of the retina, so that the reaction yield depends on the angle between the molecule and the Earth's field, not merely on field strength.

The story becomes more compelling when one moves from plausible chemistry to particular biological molecules in particular animals. "Cryptochrome" is not a single protein but a family, and birds express several cryptochrome types. The question then sharpens into something a biologist can actually test: which cryptochrome is in the right place, at the right time, in the right retinal cells, in migratory birds? And does it show the right kind of magnetic sensitivity when studied directly?

A concrete example comes from work on European robins, a species that helped shape the modern behavioral evidence for magnetic orientation. Detailed studies of robin retinas have reported a cryptochrome called CRY4 located in photoreceptor cells, including cones, with patterns that make it at least a plausible participant in a light-dependent compass. Even more suggestive is that levels of CRY4 expression in the robin retina can rise during migratory seasons. That seasonal signature is exactly what one would look for if a molecule has a role in a behavior that is itself seasonal. Biology rarely hands out proofs, but it does offer coincidences with teeth: timing, location, and necessity lining up in a way that becomes harder to dismiss as mere accident.

In 2021, a major step forward came when researchers were able to study CRY4 from a migratory songbird biochemically and ask, under controlled illumination, whether its photochemistry shows magnetic field effects consistent with the radical-pair mechanism. What makes such an experiment hard is not only the physics but the biology. Proteins do not like to be removed from their natural context. They misfold, aggregate, or lose essential partners. Demonstrating magnetic sensitivity *in vitro* demands that the protein be prepared in a functional form, illuminated appropriately, and measured with enough precision to detect small changes induced by relatively weak fields. The significance of such work is not that it instantly settles the compass question, but that it begins to bridge a gulf between behavior and molecules. "The bird turns in a field" is a behavioral statement; "this purified protein's reaction yield changes with field orientation and strength" is the beginning of a mechanistic one.

Even with that progress, the identity of the true "compass" cryptochrome remains contested, and the controversy is scientifically healthy. Another candidate, often discussed in parallel, is CRY1a, which has been reported in specific cone types associated with ultraviolet or violet sensitivity, and in patterns that suggest an ordered association with membranes in the outer segments of photoreceptors. That detail matters because order is a friend of directional sensing. A radical-pair mechanism that depends on the angle between a molecule and the geomagnetic field needs an ensemble of molecules that are not randomly oriented. Membrane-associated proteins can, in principle, be aligned by the architecture of the cell, like books on a shelf rather than leaves in a pile. The retina is full of precisely stacked membranes: the discs of rods, the lamellae of cones. If a cryptochrome were positioned in a regular way within that structure, the messy thermal environment of the cell would not automatically erase all directional information.

At this point, it helps to remember a historical lesson about biological molecules: nature often solves problems by building on old parts. A cryptochrome is not designed from scratch to be a compass. It is a blue-light photoreceptor with a well-worn role in timing and signaling. The same flavin chemistry that makes it useful for circadian regulation can, under the right constraints, generate radical pairs. Evolution does not need to invent quantum mechanics; it needs to take a molecule that already performs controlled electron transfer and tune its lifetime, its environment, and its connections so that a tiny magnetic bias becomes biologically meaningful.

The hardest part to accept, for many readers, is not that a protein can form radicals. That is everyday biochemistry. It is the suggestion that the Earth's field, about 50 μT in strength, can matter at all. Compare that to a fridge magnet, or even to the field in an MRI scanner, and it seems laughably weak. But the radical-pair idea is not about a big field doing big work. It is about a weak field nudging a probability during a brief interval when the system is exquisitely sensitive to such nudges.

A useful analogy comes from everyday decision-making rather than from magnets. A slight change in odds can matter enormously if it is applied consistently and then amplified. Casinos do not need to rig roulette by forcing the ball into a specific number every time; they just need a tiny statistical edge that persists across many spins. A radical-pair compass would be something like that: the Earth's field does not force a molecule into a different outcome, it slightly shifts the singlet—triplet balance. If a bird's retina contains many such molecules working in parallel, and if downstream biology has evolved to amplify the aggregate difference, then a faint physical effect could become a usable sensory cue.

Yet that "if" is where much of the current scientific tension lives. Even if cryptochrome is the right molecule, and even if its radical pair is magnetically sensitive in principle, how does the retina *read* the result? A tiny difference in chemical yield is not automatically a neural signal. The retina is a sophisticated piece of tissue with phototransduction cascades that can detect single photons, but those cascades have their own machinery and their own proteins. For a cryptochrome-based compass to work, the spin-dependent chemistry must couple to some biochemical pathway that can modulate ion channels, neurotransmitter release, or membrane potentials. That coupling could be direct, through protein—protein interactions, or indirect, through changes in redox state or reactive intermediates. Work mapping possible interaction partners of avian CRY4 is motivated precisely by this missing link: the molecule may be capable of a magnetic field effect, but biology must still connect that effect to perception.

There is also a deeper lesson about "hosting quantum states" in living cells. The phrase can conjure an image of a quantum computer hidden in a retina, protected from decoherence by some miraculous biological trick. The reality, as far as it can be reconstructed, is subtler and more interesting. A protein does not need to prevent decoherence in some absolute way. It needs to shape the local environment so that a particular quantum degree of freedom—here, electron spin coherence within a radical pair—survives long enough to influence a branching chemical reaction. That window might be microseconds, not seconds. It might be local and transient, not global and sustained. Biology can plausibly achieve that with ordinary tools: restricting motion, controlling distances, tuning reaction rates, and using the structural regularities of membranes.

A historical anecdote makes the stakes clearer. When the idea of magnetoreception first gained traction, some of the most convincing evidence came not from biochemistry but from behavioral experiments in which birds were placed in carefully controlled magnetic fields and their orientation tested, often by observing the direction of their scratch marks in circular cages during migratory restlessness. These experiments are, in a sense, the opposite of molecular biology: messy animals, variable motivation, and behavior filtered through statistics. Yet over decades, the behavioral signatures became crisp enough to guide the molecular search. The light dependence, in particular, functioned like a detective's clue: a constraint so strong that many otherwise reasonable mechanisms fell away. It is a reminder that good science often advances by a dialogue between levels of description. The animal points; the molecule explains.

Cryptochrome is the leading candidate not because it is the only molecule capable of radical chemistry, but because it sits at a rare intersection. It is *photoreceptive* in the right spectral range. It contains a *flavin cofactor* known to participate in electron transfer. It can form *radical pairs* as part of its natural photochemistry. It is *present in the retina*, in cell types that plausibly could contribute to a directional sense. And it belongs to a family of proteins already used by organisms to connect light to physiology, meaning it has evolutionary "wiring" for signaling.

None of this guarantees that cryptochrome is the compass, or that there is only one compass molecule, or even that birds rely on a single magnetic mechanism. Animals are rarely monogamous in their solutions. A magnetite-based sensor might provide information about field intensity or serve as a backup cue, while a cryptochrome-based mechanism provides directional (inclination) information under appropriate light. Or different bird species might emphasize different molecular pathways. The goal is not to pick a favorite story; it is to understand what kind of physical effect biology can realistically harness.

In the end, cryptochrome's candidacy is powerful because it makes the extraordinary feel almost ordinary. Take a familiar ingredient of life—a protein that uses blue light to shuffle electrons—and place it in the layered membranes of a retina. Allow those electrons, for a brief moment, to remember how they were born together, their spins correlated in a way that chemistry cares about. Let the Earth's field tip the balance ever so slightly. If a nervous system can amplify that whisper into a percept, then a bird does not need a tiny bar magnet hidden in its head. It needs, instead, a molecule that can turn a photon into a radical pair, and a radical pair into a bias, and a bias into a decision to turn left or right under the stars.

7.2 The Spin Waltz

The cryptochrome story becomes truly strange only when the conversation turns from *where* the compass might sit to *how* it could possibly work. A protein in the retina can absorb blue light and shuffle electrons around—fine. Chemistry does that all the time. The leap is to claim that the Earth's magnetic field, so feeble that it does not budge a paperclip, can nevertheless bias the outcome of that chemistry. Not by pulling on iron, not by twisting a crystal, but by changing the odds of a reaction pathway through a quantum property of electrons called *spin*.

Spin is an unfortunate word because it suggests a tiny planet whirling on its axis. Nothing is literally rotating like that. Spin is better thought of as a built-in label an electron carries, a kind of intrinsic angular momentum that comes in two basic flavors when measured along a chosen axis: "up" or "down." The important fact is that spin is magnetic. An electron behaves like a minute bar magnet, with a north—south character tied to its spin state. Put an electron in a magnetic field and its energy depends, slightly, on whether its spin aligns with or against the field.

On paper this sounds like the classical compass idea again, only shrunk down. But the radical-pair mechanism is not powered by an electron being yanked around by the Earth's field. The energies involved are much too small for that. The mechanism exploits something subtler: two electron spins can start off *correlated* by the chemistry that creates them, and their correlation can evolve in time in a way that is sensitive to weak magnetic fields. That evolution does not need large energies; it needs a well-timed quantum mixing process.

The key players are *radicals*. A radical is a molecule with an unpaired electron. Most stable molecules keep electrons paired, because pairing tends to lower energy. A radical is therefore chemically reactive and often short-lived. Biochemistry uses radicals sparingly and usually under tight control, because radicals can do damage. But controlled radicals are also useful: they allow fast electron transfers and can create switch-like behavior in reactions.

When cryptochrome absorbs a photon, it can create two radicals at once: one on the flavin cofactor (FAD) and one on a nearby amino acid, often a tryptophan in an electron-transfer chain. These two radicals form what physicists call a *radical pair*. Two unpaired electrons, born together in a single photochemical event, then separate slightly in space. At that moment, their spins are not random. They are created in a specific combined quantum state.

There are only two broad classes of combined spin states for a pair of electrons. In a *singlet* state, the two spins are paired in an antiparallel way. If one is "up" along a chosen axis, the other is "down." In a *triplet* state, the spins are aligned in a parallel way, though quantum mechanics allows three related configurations (hence "triplet"). The difference matters because chemical reactions can discriminate between singlet and triplet character. Some recombination reactions are allowed easily from singlet states but are suppressed or forbidden from triplet states, and vice versa. This is not mystical; it is the consequence of conservation rules and how electron pairing must be arranged in the products.

A radical pair created in a singlet state is therefore not just "two radicals." It is two radicals that are predisposed to certain fates. If the pair stays singlet long enough, it may recombine and return to a particular chemical form. If it converts into triplet character, recombination may be blocked, leaving time for alternative chemistry: a different product, a different signaling state. The magnetic compass proposal is that the Earth's field can influence this singlet—triplet balance and thereby influence the relative yields of those products.

At first glance, this looks impossible. The Earth's field is too weak. Yet the radical pair mechanism is not asking the field to compete with thermal energy in the usual sense. Instead, it makes use of a natural *internal* magnetic environment inside the molecule. The unpaired electrons interact with nearby atomic nuclei—hydrogen, nitrogen, and others—many of which have their own nuclear spins. Those nuclear spins generate tiny local magnetic fields that are fixed by the molecule's structure. This "hyperfine" interaction acts like a built-in set of microscopic magnets that the electron spins can feel. Importantly, hyperfine interactions can drive the electron-spin pair to oscillate between singlet and triplet character over microsecond timescales. Think of it as a beat pattern in the joint spin state, set by the molecule's internal architecture.

Now the Earth's field enters as a gentle external bias that slightly alters the rate and pattern of that oscillation. It does not need to dominate. It only needs to shift the timing enough that, when chemistry finally "measures" the spin state by allowing one reaction and suppressing another, the probabilities differ measurably for different orientations of the molecule relative to the field.

Orientation is crucial. A compass must distinguish north from south, or at least distinguish *axis* and *tilt*. If every cryptochrome molecule were tumbling freely in solution with random orientations, the magnetic effect would average away. The retina helps because its architecture is ordered: membranes are stacked, proteins can be anchored, and photoreceptor outer segments are arranged like aligned forests of microscopic columns. A population of cryptochromes held in a consistent orientation could allow the magnetic field's direction to change the reaction yield in a consistent way across many molecules. The bird would not be reading a single molecule; it would be reading a statistical pattern across a vast ensemble.

The most counterintuitive part is that this is a compass without a needle. It does not point in a direction by rotating. Instead, it changes chemistry depending on the *angle* between the molecule and the field. Many radical-pair models naturally predict an *inclination compass*: the sensor responds to the tilt of field lines (downward in the northern hemisphere, upward in the southern) and cannot distinguish magnetic north from magnetic south as polar opposites. That odd prediction matches a long-standing behavioral observation: many migratory birds behave as if they sense the axis of the magnetic field rather than its polarity. Flip the field vertically and they reverse their headings, as if "down" and "up" are what matter.

A vivid way to picture the mechanism is as a fork in a chemical road. The radical pair starts at the fork, created in a particular spin state. As time passes, internal hyperfine interactions and the external geomagnetic field nudge the system back and forth between singlet-like and triplet-like character. When the chemistry commits—when the radicals either recombine or escape toward a signaling product—the route taken depends on the spin character at that instant. Rotate the molecule relative to the field, and the oscillation pattern changes. The fork favors one road slightly more often. Over millions of events, "slightly more often" becomes a reliable signal.

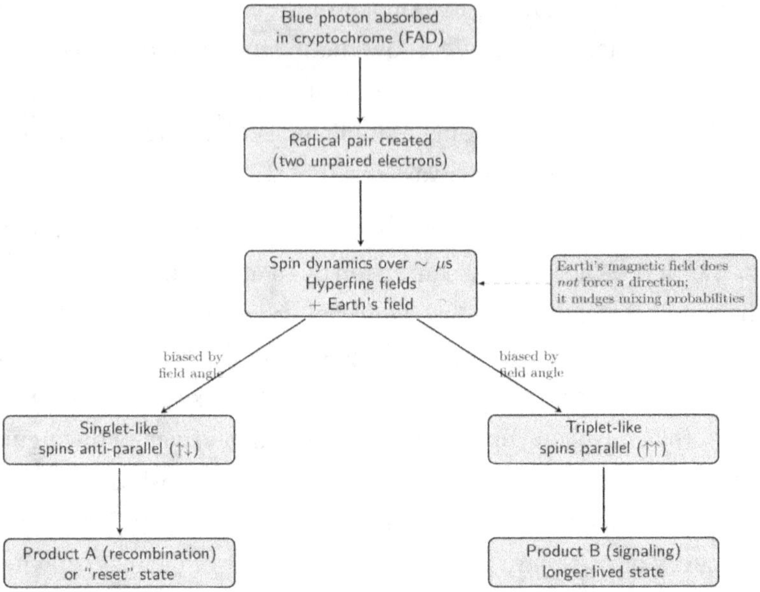

A chemical fork in the road. The same radical pair can yield different products depending on whether it is more singlet-like or triplet-like when reaction pathways "commit." The Earth's field changes the balance by a small amount, which biology can amplify by averaging over many molecules.

It is worth lingering on what that diagram does *not* say. It does not claim that the electron spins are perfectly isolated, or that they remain quantum-coherent for long times by some magical shielding. It says something more modest and more plausible: that the relevant coherence lasts long enough—microseconds rather than femtoseconds—for the spin state to influence which chemical products form. Microseconds may still sound short, but in molecular terms it can be an eternity. Many crucial chemical decisions happen on similar timescales, and cells are expert at turning small chemical biases into decisive outcomes.

How can a small bias become a sense? One answer lies in scale. The retina contains an enormous number of photoreceptor molecules, and birds are often orienting for minutes at a time. If each cryptochrome event carries a tiny directional preference in product yield, then the nervous system can integrate that information across space and time. The result need not be a crisp arrow pointing north. It may be experienced as a subtle pattern superimposed on vision: a shading, a modulation, a directional "texture" that shifts as the bird turns its head. Such phenomenology is speculative, but it is consistent with the idea of a sensory cue that is visual in its location (the eye) but not necessarily visual in the usual sense (forming an image of objects).

A historical anecdote shows how the radical pair idea moved from clever physics to biological plausibility. In the late twentieth century, the physics of spin-dependent chemistry was already well established in laboratories: chemists could measure how magnetic fields changed reaction yields in certain photochemical systems. The bold step was to suggest that evolution might have discovered the same trick. Thorsten Ritz and collaborators (2000) helped crystallize this connection by proposing a radical-pair-based compass mechanism consistent with known avian behavior. What made the proposal powerful was not that it provided a complete biological wiring diagram, but that it generated testable predictions: light dependence, an inclination compass, and susceptibility to particular kinds of oscillating magnetic fields.

Those oscillating-field predictions are rooted in a piece of physics familiar from magnetic resonance. An electron spin in a magnetic field precesses at a characteristic frequency (the Larmor frequency), which depends on the field strength. A weak radiofrequency magnetic field, if it oscillates near that frequency, can drive transitions that disturb the singlet—triplet evolution. In a controlled chemical system, such perturbations can change yields. If a bird's compass depends on radical-pair spin dynamics, then applying a weak oscillating field in the right frequency range might scramble the signal.

That argument has an almost cinematic quality: instead of blinding a bird with a bright light or confusing it with a physical magnet, one could "jam" its compass remotely with radio waves, like interfering with a broadcast. The first time this idea was tested behaviorally, the logic was stark. If the bird has a magnetite needle, a weak radiofrequency field should do essentially nothing; the forces are too small. If the bird's compass depends on delicate spin dynamics, the right oscillating field could matter even at low amplitudes. Experiments in the early 2000s, notably by Ritz and colleagues (2004), reported precisely this kind of disruption in migratory orientation under weak radiofrequency fields. It was a striking hint that the compass is not simply a mechanical sensor.

Even here, one must be careful. Biology is rarely as clean as a physics thought experiment. The precise quantitative relationship between radiofrequency disruption and radical-pair parameters remains a topic of debate, and not every experimental result lines up neatly with the simplest models. That tension is not a weakness; it is a sign that the problem is real. It forces attention onto details that matter: the lifetimes of radical pairs, the strengths of hyperfine interactions, the degree of molecular alignment in the retina, and the downstream amplification steps. A mechanism that works on a blackboard must still survive the messy constraints of proteins, membranes, and neurons.

One of the more illuminating points, though, is that the mechanism does not demand a single, unique radical pair with fixed properties. Cryptochromes can form multiple radical pairs along a chain of electron transfers, and different radical pairs can have different separations, different lifetimes, and different hyperfine environments. Work on avian CRY4 has suggested that there may be several sequential radical-pair states, not one. That multiplicity can look like a complication, but it may be a feature. Biology often uses cascades and intermediate states to tune sensitivity and timing. A short-lived radical pair might act as an initiator; a longer-lived one might be the true "sensor" state where magnetic modulation has time to accumulate; another might be optimized for signaling to partner proteins.

The "spin waltz" metaphor earns its keep here because the key phenomenon is temporal. The spins do not simply sit in singlet or triplet form like static labels. They evolve. They mix. Their joint character oscillates under the influence of internal and external magnetic fields. The chemical reaction acts as both choreographer and judge: it sets the initial state and it decides, at some moment, which pathway is taken. The magnetic field is not the dancer; it is the slight change in music that makes one step more likely than another.

There is a final subtlety that often gets lost in popular retellings. When people hear "entanglement" or "quantum coherence," they imagine fragility so extreme that warm biology seems hopeless. But spin coherence in radical pairs is not the same as keeping a delicate superposition of large objects. It is closer to what chemists routinely handle when they do electron paramagnetic resonance on radicals at room temperature. The challenge is not that coherence is impossible; it is that it must persist in a specific molecular context long enough to influence a branching reaction and, crucially, that the molecule must be arranged so that orientation information is not washed out.

In a bird's life, all of this physics must run quietly in the background while it does more urgent things: avoid predators, find food, keep warm. A molecular compass cannot be finicky. It cannot require laboratory conditions. It has to work on cloudy evenings and clear nights, in forests and over seas, in bodies that are tired and hungry. That pragmatic pressure is, in an odd way, comforting. It suggests that if the radical pair mechanism truly underlies avian navigation, it must be implemented not as a brittle quantum trick, but as a robust biochemical module: a tiny bias repeated endlessly, averaged over many molecules, and translated into neural language by the same evolutionary pragmatism that built eyes, wings, and instincts.

A compass built from spin does not pull north; it lets north *tilt* chemistry. The bird's brain, like any brain, is an engine for extracting meaning from biased, noisy signals. Give it a faint, angle-dependent chemical asymmetry in the retina, and it can do what nervous systems do best: turn probability into behavior.

7.3 Jamming the Signal

A compass built from electron spins has an awkward property: it should be easy to break. Not in the sense of snapping a needle in half, but in the sense of losing the very subtle timing that makes the mechanism work. That vulnerability is not an embarrassing weakness. It is the kind of weakness experimentalists dream of, because it turns a beautiful idea into something you can poke.

The guiding intuition is simple enough to explain at a dinner table. If a bird navigates by tiny magnets, then you would have to apply a strong magnetic force to confuse it. If it navigates by a delicate quantum process—two electron spins briefly keeping track of each other as chemistry runs—then a weak, well-chosen disturbance might be enough. The disturbance does not need to be strong; it needs to be *relevant*. A radio-frequency field can be relevant because electron spins respond to oscillating magnetic fields in the way a swing responds to well-timed pushes. One push hardly matters. A sequence of pushes at the right rhythm can change everything.

That rhythm is set by a spin's precession in a magnetic field. "Precession" is the slow wobble of a spin's magnetic axis around the direction of an external field. It is the same physics used in magnetic resonance, from chemistry labs to hospital MRI scanners, only in MRI the fields are immense and the signals are detected with carefully engineered coils. In the geomagnetic field the precession frequency is in the radio range, on the order of a megahertz for an electron. If a radical pair in cryptochrome is mixing between singlet-like and triplet-like character on microsecond timescales, then a weak oscillating field near the relevant frequencies could scramble that mixing. In effect, it would blur the difference between "more singlet" and "more triplet" at the moment the chemistry commits to one product or another.

This is why "jamming" is such an apt word. It does not imply overpowering the sensor with brute force. It implies injecting noise into a channel whose message is subtle.

One of the best-known early demonstrations of this logic came from experiments with migratory birds tested in carefully controlled magnetic conditions. The birds were not asked to fly across Europe; instead they were placed in an orientation cage, a circular arena that allows researchers to infer preferred direction from the animal's movements during migratory restlessness. This restlessness has a name, *Zugunruhe*, and it is one of those wonderfully specific German terms that science borrows when it has no equally compact English equivalent. A bird ready to migrate will hop and flutter in the direction it "wants" to go, even if it is indoors. Over many trials and many birds, those movements create a statistically meaningful compass rose.

The essential twist was to expose the birds not only to a static magnetic field (the Earth's field, or a laboratory-generated equivalent) but also to a very weak oscillating field in the radio-frequency range. Ritz and collaborators (2004) reported that such oscillating fields could disrupt the birds' magnetic orientation. That result mattered because it matched, in spirit, the radical pair idea: a weak radio field should be able to perturb a spin-dependent chemical compass, while a simple magnetite needle would not be expected to notice.

The romance of the experiment lies in the mismatch of scale. The radio-frequency field amplitudes used were far smaller than what people associate with the word "magnet." It is unsettling to think that the same kind of electromagnetic background produced by electronics could, under some conditions, be biologically relevant. It raises the possibility that an animal's sensory world includes a layer of physical structure humans mostly ignore: faint time-varying magnetic fields, usually dismissed as technologically irrelevant noise.

That possibility moved from intriguing to unavoidable with a striking set of experiments on European robins in the early 2010s. Sabine Engels and colleagues (2014) worked with birds in an urban environment and noticed something odd: under certain laboratory conditions, the robins seemed unable to orient properly with their magnetic compass. The failure was not subtle; the birds' headings became scattered, as if the compass had gone blind. Yet when the experimental setup was moved into an electrically shielded environment, or when time-varying electromagnetic noise was attenuated through grounding and shielding measures, the birds' orientation returned. Then, when noise was deliberately reintroduced, the compass failed again.

This is the experimentalist's dream: a reversible loss of function tied to a controllable physical parameter. No surgery. No drugs. No genetic manipulation. Just an invisible field that can be turned down and up, with behavior following along.

The implication is not that cities are inherently hostile to birds because they are "quantum." The implication is more precise and more unsettling: if the avian compass relies on a radical pair process, then weak broadband electromagnetic noise in the kilohertz-to-megahertz range can interfere, because that noise overlaps the timescales and frequencies relevant to spin dynamics. A field does not have to be tuned to a single frequency to cause trouble; broadband noise can act like a fog, smearing out the coherent oscillations that otherwise allow the radical pair's singlet—triplet character to carry information about direction.

At this point, a healthy skeptic asks an equally healthy question: if the compass is so fragile, how do birds navigate at all? The Earth is not an electromagnetically silent planet. Lightning strikes generate radio pulses. The ionosphere crackles. Even away from cities, natural electromagnetic noise exists. The answer may be that biology has tuned the compass to operate within a "functional window" of conditions that are usually met in the animal's typical environment, and that anthropogenic noise can push the system outside that window. It is also possible that birds combine multiple cues: stars, polarized light patterns at twilight, odors, landmarks, and magnetic information. If one channel becomes unreliable, the animal can fall back on others, much as a human driver who loses GPS can still find the way using road signs and memory.

The most persuasive aspect of radio-frequency jamming is that it is so hard to reconcile with purely classical torque-based sensors. A magnetite particle embedded in tissue might respond to changes in a static field or to strong low-frequency fields that physically torque it, but a weak oscillating field in the megahertz range does not easily translate into a mechanical force. For the radical pair mechanism, by contrast, the megahertz range is exactly where one expects sensitivity, because it matches electron spin precession frequencies in Earth-strength fields. The jamming experiments are therefore often described as a "smoking gun" for spin chemistry in living navigation.

Yet even smoking guns can misfire, and the history of science is full of results that seemed definitive until someone measured a little more carefully. The radical pair interpretation has faced a serious quantitative challenge: some theoretical and computational analyses have argued that the magnitudes of radio-frequency fields reported to disrupt orientation are difficult to reproduce with the simplest radical-pair models, given reasonable assumptions about radical lifetimes and hyperfine interactions. If a radical pair decoheres too quickly, it cannot be very sensitive. If it stays coherent for a very long time, it becomes sensitive but then must somehow avoid other sources of noise. Either way, there is a balancing act between plausibility and performance.

This is not a mere technical quibble. It forces a more refined view of what "jamming" means biologically. Perhaps the actual radical pair in cryptochrome is not the standard textbook example but a particular pair with unusual properties: a separation that reduces certain interactions, a local environment that extends coherence, or a sequence of radical pairs that effectively prolongs sensitivity. Perhaps downstream biochemical steps amplify tiny spin-dependent differences so that a field that causes a minuscule change in chemical yield still produces a significant behavioral shift. Biological amplification is not hypothetical; it is the essence of sensory transduction. A single photon can trigger a measurable response in a rod cell because the phototransduction cascade amplifies one molecular event into many ion-channel changes. If the retina can amplify a photon, it is not absurd to imagine it amplifying a small bias in cryptochrome chemistry.

Amplification, though, is a double-edged sword. The more you amplify, the more you risk amplifying noise and false positives. That makes the shielding experiments psychologically compelling: they suggest that some kinds of environmental electromagnetic noise are indeed being picked up by the compass pathway, whatever its molecular details, and are being amplified enough to matter for behavior.

There is another subtlety that strengthens the radical pair interpretation: radio-frequency fields seem to disrupt orientation without obviously disrupting vision. If the effect were simply that the radio field was heating tissue, or altering general neuronal excitability, one would expect broad behavioral disturbances. Instead, the reports describe a specific loss of compass orientation under conditions where the birds otherwise behave normally. That selectivity fits better with interference in a specialized sensory channel than with a general physiological stress.

A further twist comes from work by Roswitha and Wolfgang Wiltschko and others who have explored how birds respond to a variety of magnetic manipulations, including oscillating fields. Their studies emphasize that birds do not simply "adapt" to radio-frequency disruption in the way they can sometimes adapt to certain static field changes. A static field shifted in intensity can move the system within its functional window; the bird may recalibrate. Time-varying fields, especially in the relevant frequency ranges, can act more like sabotage, because they interfere with the primary physical process rather than simply changing the value of the parameter being sensed.

All of this would remain a fascinating story of animal behavior and environmental physics even if cryptochrome were not involved. But cryptochrome is hard to avoid once one takes the light dependence seriously. A particularly telling set of observations is that radio-frequency fields do not always seem to disrupt cryptochrome photochemistry in the simplistic way one might imagine—as if they "turn the protein off." Some experiments and analyses suggest that the effect is not merely blocking the photoreduction steps of cryptochrome, but rather perturbing the spin dynamics associated with radical pairs. That distinction matters. It implies the jamming is not a general biochemical poisoning but an interference with the very quantum-sensitive part of the process.

A concrete way to understand this is to think of a radio station. If you cut the station's power, the broadcast stops. That is like blocking photochemistry: no radicals, no signal. If you broadcast on the same frequency with static, the station is still transmitting, but the listener cannot decode the message. That is like perturbing spin dynamics: the chemistry still runs, but the directional information fails to emerge cleanly.

The modern research challenge is to connect these behavioral jamming effects to specific molecular and cellular steps. One reason this has been difficult is that cryptochrome's putative compass function likely depends on interactions with other proteins: anchors that orient it in the retina, partners that sense its redox state, signaling components that translate chemical yields into neuronal activity. Work mapping potential interaction partners for avian CRY4 has begun to populate this missing middle layer, suggesting routes by which a spin-dependent chemical difference might couple to phototransduction-related pathways or ion-channel regulation. None of these routes is yet a settled "wire diagram," but they shift the discussion from "could a protein be sensitive?" to "how might a cell read it?"

The jamming experiments also carry an uncomfortable societal implication. If anthropogenic electromagnetic noise can disrupt magnetic orientation in birds under controlled conditions, then the human-built electromagnetic environment may be an invisible pollutant for species that depend on magnetic cues. The magnitude of real-world impact is still an active topic, because wild navigation is complex and animals rarely depend on a single cue. But it is hard not to feel a chill at the thought that a migratory animal's sensory system might be vulnerable to the unintended byproducts of modern infrastructure.

This vulnerability is not a sign that the mechanism is implausible. On the contrary, it is what makes the mechanism testable. A robust magnetite needle would be stubborn; it would take a strong intervention to fool it, and strong interventions often come with confounding side effects. A spin-based sensor is, by its nature, exquisitely sensitive to certain kinds of time-varying magnetic fields, and that sensitivity gives experimenters a clean handle.

The most honest view of the evidence is therefore a blend of confidence and restraint. Confidence, because weak radio-frequency and broadband electromagnetic fields can disrupt magnetic compass behavior in ways that align naturally with a radical-pair picture. Restraint, because translating those behavioral disruptions into precise molecular parameters remains difficult, and because theoretical debates about field strengths and coherence times have not been fully resolved. Between those poles lies the real excitement: a living system may be using quantum spin dynamics not as a laboratory curiosity but as a practical sense, and the easiest way to glimpse it is to interfere gently, with nothing more than an oscillating field, and watch a bird's certainty dissolve into scatter.

Chapter 8

Reality Check

Is life truly quantum, or are we simply searching for magic in the machinery? By confronting the hostile 'warm, wet, and noisy' environment of the cell, this chapter applies the rigorous blade of Ockham's Razor to separate genuine quantum anomalies from complex, yet classical, chemistry.

8.1 The Tyranny of Thermodynamics

Walk into a modern quantum-computing laboratory and you will notice a peculiar aesthetic: stainless steel, vacuum flanges, cables carefully routed like the veins of a machine, and—at the center—something that looks less like a computer than a high-end refrigerator designed by a jeweler. That refrigerator is not a luxury. It is the point. Many of the most celebrated quantum devices only behave as advertised when chilled to temperatures a fraction of a degree above absolute zero. At those temperatures, thermal agitation quiets down. Random molecular jostling retreats. The world, briefly, becomes calm enough for delicate quantum states to remain intact.

Now compare that with the inside of a living cell. It is warm, wet, crowded, and relentlessly busy. Proteins bump. Water molecules rearrange. Ions rattle in and out of binding pockets. Membranes undulate. Even structures we draw as tidy cartoons-an enzyme active site, a pigment-protein complex-are constantly vibrating. Life is not just complicated; it is *thermally alive.* That contrast, between the cryogenic discipline of a quantum lab and the molecular mosh pit of biology, is the first and most stubborn objection to quantum biology.

The objection has a name that sounds almost like an insult: the "warm, wet, and noisy" problem. It is not a philosophical complaint about biology's untidiness. It is a quantitative claim about time. The key word is *decoherence*: the process by which a quantum system loses the ability to show interference-the hallmark of being in more than one state at once, in a way that can later recombine. Decoherence does not necessarily destroy energy or even stop dynamics. It destroys *phase information*, the invisible bookkeeping that allows quantum amplitudes to add and cancel like ripples on a pond. Without phase, a quantum superposition becomes, for all practical purposes, an ordinary classical mixture: not "both," but "either-or" with probabilities.

Thermodynamics enters because the cell is not a quiet box. It is a warm bath of many degrees of freedom-an "environment" in the physicist's sense. Any quantum system inside it is constantly nudged, monitored, and inadvertently measured by its surroundings. Each collision with a water molecule, each fluctuating electric field from a nearby ion, each vibrational twitch of the protein scaffold can imprint a little information about the system into the environment. The environment does not keep that information politely filed away. It disperses it into countless microscopic correlations, effectively unrecoverable. From the point of view of the quantum system, the phases have leaked out into the world.

It helps to translate this into something like everyday experience. Imagine trying to keep two metronomes on a table ticking in perfect step. If the room is silent and the table steady, their synchronized ticking can persist for a long time. Now put them on a table in a crowded subway station. Each footfall and vibration nudges them. Very quickly, the clicks drift out of step. The metronomes still tick; the problem is that their *relative phase* becomes unpredictable. Quantum coherence is like that relative phase, except it is far more sensitive because the "ticks" are not loud macroscopic motions but tiny molecular degrees of freedom coupled to an environment with astronomically many ways to scramble them.

A cell's thermal energy scale is set by $k_B T$, where k_B is Boltzmann's constant and T is temperature in kelvin. At human body temperature, $T \approx 310\,\mathrm{K}$, the product $k_B T$ corresponds to an energy of roughly 4×10^{-21} J, or about 25 millielectronvolts. Those numbers look small until you remember that many molecular energy differences relevant to biology-conformations, binding energies, vibrational quanta-sit in that neighborhood. Thermal fluctuations at room temperature are not a mild background hum; they are comparable to the energetic "gaps" that would otherwise keep states distinct. So the environment not only shakes the system; it shakes it *at the right scale* to matter.

There is an important subtlety here. Quantum mechanics, of course, governs all matter. No one doubts that chemical bonds are quantum, or that electrons occupy quantized orbitals, or that the stability of molecules makes no sense without quantum rules. The warm, wet, and noisy objection is aimed at *nontrivial* quantum features-superpositions that span distinguishable states, entanglement between components, coherent wave-like transport-persisting long enough to be used as part of a biological function. The tyrant is not quantum mechanics. The tyrant is *thermodynamics*, which insists that any small system strongly coupled to a hot bath will quickly forget phase relationships unless it is unusually protected.

One historical anecdote shows how deep this skepticism runs. In the 1990s, long before "quantum biology" became a fashionable label, some theorists speculated about quantum coherence playing a role in the brain-ideas involving microtubules and other structures acting as quantum information carriers. In 2000, Max Tegmark published a now-famous analysis asking a question that sounds almost rude in its simplicity: if a neuron or a microtubule were really in a quantum superposition of two different states, how long would it take the warm brain environment to decohere it? His estimates, using reasonable couplings to surrounding charges and dipoles, produced decoherence times fantastically shorter than neural timescales-so short that, compared to the milliseconds of brain signaling, the coherence would be gone essentially instantly. Whatever one thinks of the broader debate, Tegmark's calculation became a template: identify the putative quantum degree of freedom, estimate how strongly the environment can "learn" about it, and compare the decoherence time to the time the biology would need to use it.

That comparison of timescales is the reality check thermodynamics demands. For a quantum effect to matter biologically, the system must *do something* with coherence before the environment turns it into classical noise. If decoherence happens in 10^{-13} seconds but the biological step takes 10^{-3} seconds, there is a gap of ten orders of magnitude. No amount of poetic language bridges it.

Decoherence also differs from a more familiar enemy: *dissipation*. Dissipation is about energy: systems relax, heat flows, excited states decay. Decoherence can happen even when energy is conserved, because phase can be randomized without changing populations. A molecule might remain excited while losing the ability to interfere with an alternative excitation pathway. This matters because many biological processes are fast and energy-efficient; they can avoid wasting energy while still being pummeled by phase-randomizing noise.

A vivid way to picture the thermodynamic problem is to ask what it takes to build a quantum device at all. Consider a superconducting qubit. Engineers go to heroic lengths to isolate it from electromagnetic interference, from stray photons, from vibrations, from defects in materials that behave like tiny fluctuating spins. Even then, the "coherence time" of the qubit-the window during which it behaves quantum mechanically in a useful way-can be microseconds to milliseconds, depending on architecture and era. That sounds long until you realize the clock speed is also fast, so those microseconds are precious but not infinite. The entire field is a war against the environment.

Cells cannot declare war on their environment. The environment is their habitat. Water is not a contaminant; it is the medium that lets biomolecules fold, diffuse, bind, and react. Temperature is not a nuisance; it is the thermal energy that helps systems explore configurations and cross barriers. Noise is not merely a threat; it is part of how biology samples, searches, and adapts. The cell is an open system by design. So if a quantum resource is to be functionally relevant, it must either be remarkably short-lived but still effective, or be protected by some special mechanism-structural, dynamical, or chemical-that slows decoherence enough to matter.

This is where the tyranny becomes more interesting than a simple veto. Thermodynamics does not say "never." It says "pay the price." The price is usually expressed in one of two currencies: *timescale* and *lengthscale*. Superpositions that differ only slightly-say, two electron locations separated by a fraction of an angstrom, or two spin states that produce nearly identical electric fields-are harder for the environment to distinguish, and therefore decohere more slowly. Likewise, quantum dynamics that complete in tens or hundreds of femtoseconds may outrun the worst of thermal scrambling. The smaller and faster you keep your quantum trick, the less the environment can pry it apart.

That logic already suggests why certain candidate areas of quantum biology keep resurfacing. Electron and proton tunneling involve very small distances. Spin dynamics in radical pairs involve internal degrees of freedom that are relatively "quiet" electrically. Excitation energy transfer in tightly packed pigment networks can occur on ultrafast timescales. In each case, the claim is not that the cell behaves like a quantum computer. The claim is that certain sub-molecular variables may, briefly and locally, retain quantum character long enough for chemistry to exploit.

But skepticism remains because biology is full of classical look-alikes, and because thermal noise can masquerade as sophistication. A biological system can appear "finely tuned" simply because evolution has found robust classical mechanisms that work under noisy conditions. If a process is already fast and efficient classically, thermodynamics invites a blunt question: why invoke fragile coherence at all? The simplest answer, often, is that you should not.

The physics behind this bluntness is clearer if we translate decoherence into a story about information. A quantum superposition is not just "two possibilities." It is a specific *relation* between possibilities-amplitudes with phases. The environment does not need to measure the system in the dramatic sense of a laboratory apparatus with a readout. It only needs to interact in a way that correlates distinct system states with distinct environmental states. If the environment ends up in $|E_1\rangle$ when the system is in $|1\rangle$, and in $|E_2\rangle$ when the system is in $|2\rangle$, then the joint state becomes

$$\alpha\,|1\rangle|E_1\rangle + \beta\,|2\rangle|E_2\rangle.$$

If $|E_1\rangle$ and $|E_2\rangle$ are effectively distinguishable-orthogonal, or close enough given the environment's complexity-then interference between $|1\rangle$ and $|2\rangle$ becomes inaccessible to any observer who does not also control the environment. In a cell, no one controls the environment. The environment is, in effect, a busy gossip: it broadcasts which alternative happened, scattering the evidence into countless degrees of freedom.

Crowding makes this worse. Cells are not dilute solutions. They are packed with macromolecules. That packing increases the frequency and variety of interactions. It also means local electric fields fluctuate strongly, because charged residues and ions are everywhere. Even a neutral molecule carries partial charges and polarizability; it is sensitive to the jiggling electrostatic landscape. If your proposed quantum variable changes the charge distribution even slightly, the environment can "notice" quickly.

Water, in particular, is a formidable decoherer. It is polar, it forms hydrogen-bond networks that rearrange on ultrafast timescales, and it provides an enormous density of vibrational and rotational modes. Put a quantum system in water at room temperature and you have placed it into intimate contact with a bath that offers countless channels for phase randomization. This is the deep reason the phrase "warm and wet" became shorthand for skepticism.

And yet, life persists precisely because it can function under that chaos. Thermodynamics, after all, is not only an enemy. A cell maintains order by exporting entropy, using metabolic energy to hold itself far from equilibrium. That far-from-equilibrium drive gives biology a kind of dynamical cleverness: it can create structured microenvironments, protect sensitive intermediates, and channel energy flow along preferred pathways. None of that guarantees quantum coherence. But it does mean biology has tools that inert matter lacks.

One tool is architectural: bury key components inside proteins. A pigment molecule in a photosynthetic complex is not floating freely in water; it is held in a pocket, surrounded by amino acids that both tune its energies and buffer its fluctuations. Another tool is dynamical: rapid processes can occur before the environment fully scrambles phases. A third tool is statistical: even if individual quantum events are fleeting, their aggregate effect might bias a chemical outcome in a way selection can exploit.

Still, the tyranny of thermodynamics demands humility. If someone claims that a biological system maintains large-scale, long-lived superpositions-say, a protein being in two distinct shapes at once over milliseconds in warm water-the burden of proof is crushingly high. Not because biology is unimaginative, but because the environment is unimaginative in a brutally consistent way: it couples, it correlates, it decoheres.

The most honest way to treat the warm, wet, and noisy objection is to make it concrete. Whenever a claim of "quantum advantage" in biology appears, the first questions are not mystical. They are almost mundane: How different are the competing states in space, charge distribution, or magnetism? How strongly do they couple to surrounding water, ions, and vibrations? How quickly does the biological process need the coherence to last? If the numbers do not line up, thermodynamics wins.

And thermodynamics wins often enough that the rare counterexamples-if they exist-become all the more compelling, not because they violate physics, but because they manage to live with it. They would be small victories carved out of a hot, noisy world: not a repeal of the second law, but a clever exploitation of the narrow cracks it leaves open.

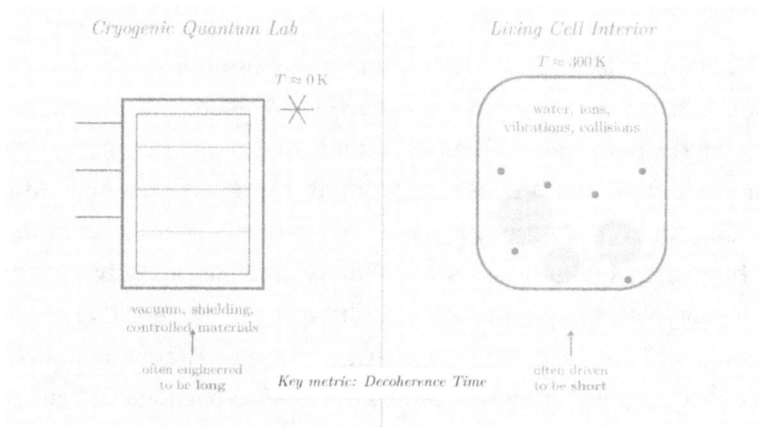

Thermodynamics can feel like a scolding: stop dreaming, everything decoheres. But it is better understood as a demand for precision. If a quantum effect is claimed in biology, the claim must come with a plausible answer to a brutally simple question: how does it survive contact with a hot, wet environment long enough to make a difference? In that demand, there is a kind of respect. The cell is not being dismissed as too messy for physics; it is being taken seriously as a physical system—one whose relentless thermal motion sets the rules of what can and cannot remain coherent in the midst of life's necessary noise.

8.2 Ockham's Razor and Quantum Ghosts

William of Ockham, a fourteenth-century English friar, did not invent a law of nature. He offered a habit of mind. When several explanations account for what you see, prefer the one that multiplies assumptions the least. This "razor" has a certain moral appeal: don't clutter the world with invisible machinery unless you are forced to. Science, at its best, is a disciplined refusal to be impressed by your own cleverness.

Quantum biology is a perfect arena for that discipline because quantum mechanics arrives with a seductive advantage. It sounds deep. It sounds like a hidden layer. It sounds like an answer that must be there, somewhere, because the world is ultimately quantum. In the wrong hands, that turns into a kind of intellectual permission slip: if a biological system looks astonishingly efficient or oddly sensitive, why not say it is exploiting superposition or entanglement? The problem is not that such claims are always wrong. The problem is that biology supplies an endless supply of classical phenomena that look, at a distance, like quantum wizardry.

A helpful starting point is to separate three meanings of "quantum," which are often mixed together in popular discussions.

First, there is *quantum as bookkeeping*: the fact that chemistry works at all because electrons occupy quantized states and bonds come from the quantum rules of indistinguishable particles. This is not controversial; it is as universal as the periodic table.

Second, there is *quantum as mechanism*: tunneling, zero-point motion, quantized vibrations, spin selection rules. These are still ordinary parts of physical chemistry, but they can have functional consequences in enzymes, electron transfer chains, and photochemistry. They do not require fragile, long-lived coherence spread across many possibilities.

Third, there is *quantum as resource*: coherent superpositions, interference, entanglement, nonclassical correlations that cannot be mimicked by a classical probabilistic model without sneaking in equivalent complexity. This last category is what makes quantum computing sound revolutionary and what makes quantum biology sound, to many physicists, like a dare.

Ockham's razor bites most sharply when someone slides from the first two meanings into the third without noticing. The mere fact that a biological process uses quantum rules (all chemistry does) does not mean it harnesses uniquely quantum resources in a way that changes function. A protein does not become a quantum computer just because electrons are quantum. Calling that "quantum biology" would be like calling plumbing "quantum hydrodynamics" because water molecules obey the Schrödinger equation.

The danger is not only exaggeration; it is misdiagnosis. Complex classical systems can generate signatures that resemble quantum coherence, especially when you observe them with sophisticated tools. In biology, you rarely observe a single isolated molecule. You observe ensembles: many copies, many slightly different environments, many pathways overlapping in time. If you see oscillations, correlations, or surprisingly robust performance, it is tempting to interpret them as evidence that nature has found a quantum trick. But the world is full of oscillations. A violin string and a bridge both ring. A crowd sways. A protein vibrates. The presence of oscillatory behavior does not, by itself, diagnose quantum coherence any more than the presence of a heartbeat diagnoses a particular emotion.

The phrase "quantum ghosts" captures a specific kind of mistake: treating something that *looks* quantum as if it must *be* quantum, when an unromantic classical explanation is already sitting nearby. The ghost is the impression of a quantum entity—coherence, entanglement, wave-like motion—haunting a dataset that could have been produced by ordinary physics plus biological complexity.

A concrete example comes from the most celebrated arena of quantum biology: photosynthesis. Certain light-harvesting systems transfer excitation energy with remarkable efficiency. When a pigment absorbs a photon, the energy must move to a reaction center before it is lost to fluorescence or heat. In 2007, Gregory Engel's lab used two-dimensional electronic spectroscopy—a technique that can correlate excitation and detection frequencies on femtosecond timescales—to report oscillatory "beating" signals in the Fenna—Matthews—Olson (FMO) complex, a pigment-protein structure used by green sulfur bacteria. Those oscillations were widely interpreted as evidence that electronic excitations were moving coherently among pigments, like a wave exploring multiple routes at once.

The excitement was understandable. If the energy really moved by quantum interference, it would offer a vivid counterpoint to the warm, wet, and noisy skepticism. But within a few years, a sober complication emerged: vibrational motions of molecules can also produce oscillatory signals in the same experiments. The protein scaffold and the pigments are not rigid. They support quantized vibrations, and those vibrations can be coherently excited by ultrafast laser pulses. Even if the electronic excitation is not coherently delocalized for very long, the nuclear motions can create "beats" that look similar in the data. Vivek Tiwari and colleagues (2013) showed that vibrational wavepackets, including in the electronic ground state, can generate long-lived oscillatory features that might be mistaken for electronic coherence. The result was not that quantum mechanics disappears from the story; vibrations are quantum too. The result was that the particular *resource* being claimed—long-lived electronic coherence doing functional work—was harder to pin down than the first wave of headlines suggested.

Ockham's razor enters here in a subtle way. Suppose an experiment shows oscillations. You can propose Model A: electronic coherence among pigments persists for hundreds of femtoseconds and improves energy transport. Or Model B: the observed oscillations come largely from vibronic and vibrational effects that do not require sustained electronic coherence, while energy transfer can be explained by a mixture of coherent and incoherent processes or even largely incoherent hopping influenced by the protein environment. If both models can reproduce the data, the razor does not declare Model B "true." It asks you to be cautious about claiming Model A's extra entities. In practice, it asks for new predictions that distinguish them, and for analyses that punish models for being too flexible.

Flexibility is the hidden enemy of interpretation. A model with many adjustable parameters can often fit almost anything. In the language of statistics, it overfits. In the language of everyday life, it tells a good story no matter what happens. Quantum models can become overfit in a particularly elegant way because quantum dynamics live in complex amplitudes, not just probabilities. Phases, couplings, dephasing rates, spectral densities—there are many knobs, and if you allow yourself to tune them freely, you can often make a simulated spectrum look plausibly like an experimental one. Open-quantum-system modeling is valuable precisely because it forces you to say, explicitly, what the system is, what the environment is, and how they couple. It also forces you to confront what you do *not* know. But it does not magically prevent overfitting. It simply makes the assumptions legible.

There is a second complication that Ockham's razor helps to diagnose: the confusion between *existence* and *relevance*. It is entirely possible for a biological system to exhibit a fleeting quantum coherence that leaves a measurable trace, and yet for that coherence to be irrelevant to the biological function. A camera flash can freeze the motion of a hummingbird's wing, but the bird is not using the flash. Likewise, a femtosecond spectroscopy experiment might reveal coherent dynamics that occur because molecules are quantum objects under ultrafast excitation, not because evolution has tuned those dynamics for a functional advantage.

This distinction matters because biology is filled with processes that are already "good enough." Photosynthetic energy transfer, for example, is often discussed as near-perfect. But "near-perfect" depends on what you define as success. If the relevant comparison is between transfer time (picoseconds) and fluorescence lifetime (nanoseconds), then even an incoherent transport model can yield very high efficiency simply because the energy usually arrives long before it can leak away. David Wilkins and Nikesh Dattani (2014) pressed this kind of point in the FMO debate: if classical or incoherent models already explain the efficiency and timescales, coherence may not be a necessary explanatory ingredient, even if some coherence exists.

That is the razor in action: do not add a mechanism solely because it is exciting; add it because the data *require* it, because it improves predictive power under constraints, because it explains a discrepancy that simpler models cannot.

None of this is an argument for cynicism. It is an argument for careful criteria. What, then, would count as good evidence that a biological phenomenon is not just quantum in the trivial sense, but functionally quantum in the resource sense?

One criterion is *interference as a functional signature*. Interference is the uniquely quantum effect that turns amplitudes into outcomes in a way no classical random process can perfectly mimic without additional hidden structure. If a biological system's output depends on the relative phase between two pathways—and if changing that phase (by a controlled perturbation) changes function in a way that cannot be explained by classical kinetics—then one has a strong case.

Another criterion is *timescale alignment*. The coherence must persist long enough, and couple strongly enough, to influence the functional step. This is not a demand for long coherence times in an absolute sense. Hundreds of femtoseconds can be long if the relevant process completes in hundreds of femtoseconds. But the coherence must be commensurate with the biology, not merely detectable with an ultrafast instrument.

A third criterion is *necessity under perturbation*. If you can disrupt the purported quantum feature without otherwise breaking the system, and the function degrades in a way predicted by the quantum model but not by the best classical alternatives, that is persuasive. In practice, this is hard. Biology is not a clean circuit board. Change one thing and you often change many things. But the logic is sound: a mechanism earns its place when its removal removes the effect.

These criteria sound straightforward until you try to implement them. Biology resists clean interventions. Many experiments are done at low temperatures or in isolated complexes because the signals are easier to see. Yet the claim of functional relevance usually concerns physiological conditions. Even defining "function" can be tricky: is the goal to maximize speed, minimize energy loss, maximize robustness to disorder, or something else? A system might have coherence and yet be selected for robustness, with coherence as an accidental byproduct rather than a chosen feature.

Ockham's razor also guards against a more psychological trap: the tendency to treat "quantum" as a synonym for "mysterious." Once a process is labeled quantum, it can feel as though ordinary skepticism no longer applies, as though the correct response is awe rather than analysis. This is precisely backward. Quantum explanations should be *more* demanding, not less, because they introduce entities that are notoriously sensitive to environmental disturbance and notoriously easy to invoke without constraint.

There is a historical lesson here from an older controversy, one that has nothing to do with biology: the luminiferous aether. In the nineteenth century, light was understood as a wave, and waves were understood as vibrations of something. So physicists posited an invisible medium filling space: the aether. It was not a foolish idea; it was a reasonable inference given the conceptual tools of the time. But as experiments improved—most famously the Michelson—Morley experiment in 1887—the aether became increasingly baroque. It was supposed to be rigid enough to support fast vibrations, yet undetectable to moving matter, and it had to behave differently depending on the observer. The razor did not kill the aether by itself. Empirical failure did. But the razor was the intellectual discomfort that made physicists willing to consider that the entire scaffolding might be unnecessary. Einstein's 1905 special relativity removed the need for the aether not by denying waves, but by changing the underlying framework. The ghost vanished because a simpler, more predictive structure replaced it.

Quantum ghosts in biology are not aether in disguise, but the moral is similar. If one finds oneself continually adding invisible quantum features to rescue an explanation—entanglement here, coherence there, a specially protected subspace over there—without producing distinctive predictions that survive harsh tests, then one is not doing daring science. One is repeating the aether's late-stage pathology: an explanation that grows more elaborate precisely because it is not constrained.

Open-quantum-system theory can help keep the razor sharp. Instead of speaking vaguely about "maintaining coherence," one writes down a Hamiltonian for the system, a model for the environment, and a coupling. One specifies whether the environment causes pure dephasing (phase scrambling) or dissipation (energy loss), and on what timescales. One confronts disorder: biological complexes are not identical copies; their energies fluctuate. One checks robustness: does the effect persist under parameter variations that are realistic for messy biology, or does it require fine tuning that evolution would struggle to maintain? One compares to classical stochastic models that include the same disorder and noise. The discipline is not that quantum models are forbidden; it is that quantum models must earn their keep.

The most constructive way to apply Ockham's razor is not to shave away quantum ideas prematurely, but to sharpen the question. When a biological observation seems to "demand" quantum resources, ask what exactly is being demanded. Is it merely speed? Classical diffusion can be fast over nanometer scales. Is it efficiency? Efficiency can be high simply because losses are slow. Is it sensitivity? Classical nonlinear systems can be exquisitely sensitive. Is it a correlation? Coupled oscillators can correlate. Only when the demanded feature is *interference-dependent* does the specifically quantum explanation become difficult to replace.

A final twist makes the story more interesting: sometimes the most parsimonious explanation *is* quantum, not classical. This happens when the classical story becomes contrived. Consider electron transfer in chemistry. Before quantum tunneling was widely accepted as a routine mechanism, some classical pictures required electrons to "climb" energy barriers in ways that did not match observed temperature dependencies. The quantum description—electrons tunneling through barriers, with rates sensitive to distance and reorganization of the environment—was not an exotic add-on. It was the simplest way to make sense of the data. Ockham's razor does not favor classical explanations; it favors explanations that fit without needless invention. Quantum mechanics can be the simpler story when nature insists.

That is the intellectual posture quantum biology needs: neither credulous nor dismissive. When a claim is made, the right response is not "cells are too warm for quantum effects" as a slogan, and not "everything is quantum so anything goes" as a shrug. The right response is to ask, with friendly insistence, what the proposed quantum resource is, what it changes about the function, what alternative classical mechanisms have been ruled out, and what new prediction follows that could not have been guessed from chemistry's already formidable toolkit.

When those questions are taken seriously, the ghosts tend to behave. Some will evaporate, revealed as misread oscillations or classical dynamics wearing an impressive costume. A few may remain stubbornly solid, not because "quantum" is a magical word, but because interference, tunneling, or spin dynamics really do provide the leanest account of a biological feat. That is when the razor stops feeling like a weapon against imagination and starts feeling like what it is meant to be: a way to keep wonder tethered to reality.

8.3 Catching Lightning in a Bottle

A hummingbird's wings are not invisible. They are simply too fast for the human eye to resolve. You can stare straight at the bird and still miss the detail that matters: the precise twist of feathers that makes hovering possible. Biology has many such hidden motions. Some are slow and obvious, like a seedling bending toward a window. Others are so fast that "motion" is almost the wrong word, more like a brief rearrangement of possibility. When quantum biologists talk about coherence surviving in a living system, they are usually talking about events that live on femtosecond timescales. A femtosecond (10^{-15} s) is to a second what a second is to about 32 million years. That is the speed at which many molecules redistribute energy after absorbing light, and it is also the timescale on which thermal noise is most effective at scrambling phase.

So the experimental problem is brutal. If the cell is a noisy crowd and coherence is a whispered conversation, you do not solve the problem by listening harder. You solve it by listening faster than the noise can drown out the whisper.

The revolution that made modern quantum biology discussable was not a single discovery about a bird's compass or a bacterium's photosynthesis. It was an instrument: ultrafast spectroscopy, and in particular the family of techniques that use trains of femtosecond laser pulses to interrogate matter before it has time to forget what just happened.

The basic idea of ultrafast spectroscopy is simple enough to explain at a dinner party, which is one reason it has become a staple of popular science. You hit a sample with a very short pulse of light (the *pump*) to start a process—excite an electron, create a short-lived chemical intermediate, kick a vibrational motion into a coherent swing. Then, after a controlled delay, you hit it with a second pulse (the *probe*) and measure what changed. By varying the delay time between pump and probe, you build up a time-resolved movie of ultrafast dynamics.

But the devil is not in the concept. It is in the precision. To control delays on the order of tens of femtoseconds, you cannot use ordinary electronics. Light travels about 300 nm in one femtosecond—roughly the width of a bacterium. Changing a delay by a few femtoseconds means changing an optical path length by a few micrometers, which you can do with a carefully moved mirror on a translation stage. It is a strange kind of timekeeping: you measure time by sliding mirrors by distances smaller than the thickness of a hair.

Historically, this capability was not a gentle improvement; it was a leap. The arc of ultrafast lasers runs through the late twentieth century's breakthroughs in mode-locking, nonlinear optics, and pulse compression. In 1999, Ahmed Zewail received the Nobel Prize in Chemistry for femtochemistry—using ultrafast lasers to observe chemical reactions in real time. The cultural effect was immediate: chemistry was no longer a set of "before and after" snapshots inferred from products; it could be watched as it unfolded, at least for carefully prepared systems.

Quantum biology benefited from that cultural shift. Once a technique exists that can resolve femtosecond dynamics, it becomes plausible—almost irresistible—to ask whether biology contains functional events that occupy that fleeting window. Without such tools, the warm, wet, and noisy objection is too easy. You can always say, correctly, that coherence will be short-lived. The more interesting question is whether anything biologically meaningful happens before it dies.

Two-dimensional electronic spectroscopy (2D-ES) is a flagship example of this new experimental power. David Jonas (2003) helped formalize the method and its interpretation, but the intuition can be conveyed with a metaphor. Ordinary spectroscopy is like asking a crowd, "What notes can you sing?" and measuring the overall sound. Two-dimensional spectroscopy is like asking, "If I prompt you with this note, what note do you answer with, and how does that relationship change with time?" It correlates excitation frequency with detection frequency, producing a two-dimensional map rather than a single line. The extra dimension can reveal couplings between states, pathways of energy flow, and coherent beating patterns that would be smeared out or hidden in one-dimensional measurements.

In practice, the method uses multiple ultrafast pulses arranged in a precise sequence. The sample's response is not simply absorbed light; it is an emitted signal field generated by nonlinear interactions. One can think of the pulses as writing and reading a short-lived quantum "memory" into the sample—coherences between states—then detecting how that memory evolves. By scanning delays between pulses, 2D-ES can separate dynamics that are otherwise entangled: population transfer, dephasing, vibrational motion, spectral diffusion, and couplings between excitations.

This is where the phrase "catching lightning in a bottle" earns its keep. Lightning is not mysterious because it violates physics; it is mysterious because it is fast, bright, and embedded in turbulent air. Ultrafast spectroscopy is a way of bottling a comparable kind of event: a fleeting coherent evolution inside a strongly interacting environment.

The most famous lightning bolt in quantum biology struck in 2007. Gregory Engel's lab reported 2D-ES measurements on the Fenna–Matthews–Olson complex, performed at low temperature (around 77 K, the temperature of liquid nitrogen). They observed oscillatory features—quantum beatings—interpreted as evidence that electronic excitations in the complex retained coherence for hundreds of femtoseconds. The result did not simply add a data point. It changed the tone of the field. Quantum coherence was no longer confined to textbook thought experiments or carefully isolated atoms. It appeared, at least under some conditions, in a biological complex with a clear function: moving energy from captured light to a reaction center.

Skepticism arrived quickly, and it was healthy skepticism. Low temperature is not a cell. The crucial question was whether similar phenomena exist near physiological conditions. Follow-up work, including that of Gitt Panitchayangkoon and colleagues (2010), reported coherent signals in FMO at higher temperatures, even approaching regimes closer to those of living organisms, with coherence persisting on the order of hundreds of femtoseconds (the abstract emphasized values around ~ 300 fs). Those numbers matter because they are not abstract. Energy transfer in photosynthetic complexes often occurs on picosecond timescales, and fluorescence lifetimes are typically nanoseconds. Hundreds of femtoseconds is not long, but it is not negligible in that context.

At this point, it became tempting to tell a simple story: plants and bacteria evolved to use quantum coherence to boost efficiency. That story is easy to repeat and hard to resist. It also triggers every warning bell from Ockham's razor. Oscillations in 2D-ES are not a signed confession from nature saying "I used electronic coherence to do work." They are a signal that something coherent is happening in the response of the system to ultrafast excitation. The coherence could be electronic, vibrational, or a mixture—what chemists call *vibronic* coherence, where electronic and vibrational degrees of freedom are coupled strongly enough that the neat separation between them breaks down.

The debate over electronic versus vibrational origins of oscillations is not pedantry. It is the difference between two kinds of claims. If the coherence is primarily vibrational, then the system may still be deeply quantum (vibrations are quantized), but the functional story shifts. Vibrational coherence could simply be an accompaniment to electronic dynamics, excited by the laser pulses, without implying that electronic superpositions are long-lived or functionally exploited. If the coherence is truly electronic and persists in a way that affects energy transfer pathways, then it suggests a more direct use of quantum resources.

Work such as Tiwari, Peters, and Jonas (2013) sharpened this controversy by showing concrete mechanisms through which vibrational wavepackets—sometimes even in the electronic ground state—can create oscillatory features in 2D spectra that resemble electronic coherence. The upshot is not that the original experiments were wrong; it is that interpretation in complex molecular systems is a competitive sport. Many processes occur on similar timescales, and the laser pulses can coherently excite multiple degrees of freedom. To sort them out, one needs careful controls, isotope substitutions, temperature dependence, polarization sequences, and detailed modeling that predicts not only that oscillations exist but where they should appear in the 2D map and how they should decay.

This is where ultrafast spectroscopy becomes more than a high-speed camera. It is also a kind of interrogation technique. Different pulse sequences can emphasize different pathways. Polarization control can select certain transitions and suppress others. Comparing rephasing and non-rephasing signals can help identify whether observed features are consistent with particular coherence types. The experiment is not merely recording what happens; it is asking structured questions designed to corner the physics.

Even with such sophistication, a second problem lurks: biology is heterogeneous. Proteins fluctuate. Complexes have static disorder—slightly different pigment energies across an ensemble—and dynamic disorder—energies that shift over time as the protein environment moves. These inhomogeneities can smear out coherent features. Paradoxically, ultrafast methods can still see coherence because they are fast enough to "beat" some of the averaging, but ensemble measurements always carry the risk that what looks like a coherent oscillation is, at least in part, a rephasing artifact or an emergent feature of many slightly different systems responding similarly.

The right response is not despair; it is humility about what is being measured. Ultrafast experiments produce exquisitely information-rich data, but that richness makes them vulnerable to over-interpretation. A single time trace with a few oscillations can be made to support almost any story. The strength of 2D-ES is that it produces patterns—relationships between frequencies and times—that can discriminate between models when analyzed with restraint.

The deeper conceptual payoff of these experiments is that they force the thermodynamic objection into a more nuanced form. The warm, wet, and noisy problem remains real, but it no longer reads as "coherence cannot exist." Coherence can exist fleetingly, and nature can support it in structured environments like pigment-protein complexes. The question becomes: what does the environment do to the coherence, and can that influence be beneficial rather than purely destructive?

Here an idea that initially sounds like heresy becomes almost obvious once stated: *a little noise can help.* In 2008, Masoud Mohseni, Patrick Rebentrost, and colleagues developed models of environment-assisted quantum transport. The caricature of quantum coherence is that it is always fragile and always harmed by noise. But in disordered networks, perfectly coherent evolution can become trapped by interference, repeatedly revisiting the same sites without efficiently finding the target. Moderate dephasing can break the trapping and promote exploration, in effect blending wave-like motion with a classical-like tendency to forget phase, which can help the system avoid bad interference patterns. The result is not a free lunch. Too much noise kills coherence and yields slow hopping; too little noise risks localization or trapping. The claim is that there can be an intermediate regime where transport is optimized.

This is an appealing idea for biology because biology rarely optimizes by eliminating noise. It often optimizes by working with it. Yet the same idea illustrates why experimental access matters. It is one thing to propose that noise-assisted transport could improve efficiency; it is another to show that a real biological complex sits in that intermediate regime and that changing environmental coupling changes function in the predicted way. Ultrafast spectroscopy, paired with careful perturbations and realistic modeling, is one of the few ways to put such claims on a quantitative footing.

The most compelling virtue of ultrafast methods is that they are not restricted to photosynthesis. Any process initiated by light—vision, photoprotection, photorepair—can, in principle, be interrogated on femtosecond timescales. Even non-photochemical processes can be studied with variants of ultrafast techniques, because one can use light to trigger a change (for example, by releasing a "caged" compound) or to probe vibrational modes that report on structure.

This experimental reach also keeps the discussion honest about what counts as "quantum biology." Consider radical-pair magnetoreception, often proposed as a mechanism for birds sensing Earth's magnetic field. The core physics involves electron spins created in a correlated state by a photochemical reaction. The subsequent spin dynamics, influenced by weak magnetic fields, can affect chemical reaction yields. This is quantum in a very specific way: it depends on spin coherence and the quantum rules for singlet and triplet interconversion. It also has a clear experimental handle, because one can test magnetic field effects on chemical yields and, in some systems, perturb the spin dynamics with radiofrequency fields. It is not a claim about macroscopic superpositions; it is a claim about a delicate internal quantum degree of freedom influencing chemistry. The point is not to settle the bird debate in a single paragraph, but to note how ultrafast and spin-sensitive methods together shift the conversation from "could it be quantum?" to "what, exactly, is the quantum variable, and how do we measure it before it decoheres?"

A useful way to keep perspective is to remember how short a femtosecond really is in biological terms. Enzymes often undergo conformational changes on microsecond to millisecond timescales. Neurons fire on millisecond scales. Metabolic pathways unfold over seconds to minutes. Against that background, femtosecond coherence seems irrelevant—like arguing that the motion of individual water molecules determines the plot of a novel. But the analogy fails because some biologically crucial events happen at the moment of energy capture or charge separation, and those events can set initial conditions that cascade into slower steps. Vision begins with retinal isomerization triggered by a photon. Photosynthesis begins with creating an exciton and then separating charge. If the earliest steps are ultrafast, then even a brief coherence can bias which pathway is taken, which intermediates are formed, and how efficiently energy is routed.

Ultrafast spectroscopy has made those first steps accessible. It has also, perhaps unintentionally, made the field more mature by making it harder to bluff. When you can measure dynamics on the relevant timescale, vague claims about "long-lived coherence" stop being impressive. You must specify: which coherence, between which states, for how long, under what conditions, and with what functional consequence? The best experiments do not merely show that something oscillates; they show how that oscillation maps onto a network of couplings and how it changes with controlled perturbations.

The most striking aspect of this revolution is how it changes our sense of what is "observable." For most of scientific history, femtosecond dynamics were not merely unknown; they were, in a practical sense, nonexistent. You could not meaningfully talk about watching a molecule reorganize in 100 fs, because no instrument could timestamp events with that resolution. Once the instrument exists, the world becomes richer. It is not that nature suddenly acquired coherence. It is that human beings acquired the ability to notice it, to argue about it rigorously, and to demand that any claim about function survive the combined pressure of thermodynamics, Ockham's razor, and actual data.

A femtosecond experiment ends, inevitably, with a macroscopic trace: a spectrum on a detector, a map on a screen, a set of oscillations in time. The elegance of the method is that it translates unimaginably fast quantum motion into something slow enough to be analyzed by ordinary minds. That translation is never perfect, and it is never free of interpretive risk. But when it works, it catches something rare: the brief moment when biology, despite all its warmth and wetness, still moves quickly enough that quantum phase has not yet been completely erased, and the microscopic rules of interference can leave a visible fingerprint in the chemistry of life.

Chapter 9

Nature's Blueprint 2.0

Having uncovered the quantum machinery within living systems, humanity now stands on the brink of reverse-engineering billions of years of evolution. This chapter examines how the subtle dance of excitons and spins is inspiring a new generation of hyper-efficient solar cells, precise navigation sensors, and perhaps even a radical rethinking of pharmacology.

9.1 Harvesting Light at the Quantum Edge

A leaf has an unfair advantage over a rooftop solar panel. Not because chlorophyll is a better semiconductor than silicon, or because plants have solved the problem of wiring electricity to your toaster. The advantage is subtler: a photosynthetic organism is astonishingly good at *not wasting* an absorbed photon. In dim forest light, in shifting clouds, in water that filters the spectrum into unfamiliar colors, a bacterium or a plant antenna complex grabs energy and shepherds it to a reaction center with a reliability that engineers envy. The trick is not a single miracle ingredient but a style of design: pigments held in a soft, vibrating scaffold, with just enough disorder to explore options and just enough structure to avoid getting lost.

To translate that style into technology, it helps to be clear about what, exactly, is being moved around. When a pigment molecule absorbs a photon, it does not usually eject a free electron the way a solar cell does. Instead it creates an *exciton*: a bound package of energy shared between an excited electron and the hole it leaves behind. The exciton is not a physical particle you could scoop up; it is a quantum state spread over one molecule, and sometimes over several, which can migrate by handing its excitation to a neighbor. That migration is the first leg of photosynthesis: get the energy from wherever it was absorbed to the chemical machinery that can use it before it is lost as heat or re-emitted as light.

In a simple picture, exciton motion resembles a drunkard's walk. The excitation hops randomly from pigment to pigment, each hop nudged by thermal vibrations, until it either reaches the reaction center (success) or falls into a trap (failure): perhaps a pigment where the energy is dissipated as heat, or a pathway that leads nowhere. For decades, many scientists assumed the story ended there. Biology, after all, is warm and wet, and warm and wet is the enemy of quantum delicacy. The surprise was not that excitons move quickly—they must—but that there were hints of motion that looked less like hopping and more like a kind of coordinated glide.

The historical jolt came from spectroscopy, and it has the flavor of a detective story because the evidence is indirect. In 2007, Gregory Engel's group used a technique called two-dimensional electronic spectroscopy on a bacterial light-harvesting component known as the FMO complex. The method is a kind of ultrafast strobe-light interrogation: a sequence of laser pulses creates and probes coherent superpositions of excitonic states, and the resulting signal can reveal how electronic excitations are coupled. What they reported were oscillations—"quantum beatings"—suggestive of coherence between excitonic states. Coherence here means that the excitation cannot be described as being *either* on pigment A *or* pigment B; it is in a superposition with a definite phase relation between components. The phase is the delicate part: random interactions with the environment tend to scramble it, a process called *decoherence*.

The initial excitement was understandable. If coherence persists long enough, an exciton could, in a sense, sample multiple routes at once, biasing its energy flow toward the reaction center. That is a tempting image because it resembles a quantum walk, the quantum cousin of a random walk. But the story immediately developed a second act: careful follow-up work suggested that what the spectroscopy sees can mix electronic and vibrational coherences, and that electronic coherence may decay much faster under realistic conditions than early interpretations implied. By the mid-2010s, a more cautious consensus had emerged: some coherent dynamics are real and measurable, but their lifetimes at ambient conditions may be tens to a few hundreds of femtoseconds, not long-lived, not mystical, and not a free pass to "quantum magic." The interesting question became more engineering-friendly: can short-lived coherence, when embedded in a noisy environment, still improve performance?

That question points to one of the most productive ideas to come out of the photosynthesis debate: *environment-assisted quantum transport*. The name is clumsy but the concept is elegant. In a purely coherent system, energy can slosh back and forth, and, in disordered networks, it can even become localized—stuck by interference the way a wave can form a standing pattern that refuses to move on. In a purely incoherent hopping system, energy diffuses, but it may wander too long, increasing losses. Somewhere between these extremes lies a regime where a modest amount of environmental noise actually helps: it shakes the system just enough to prevent localization and to keep the excitation exploring, while retaining enough coherence to exploit wave-like shortcuts. Noise becomes a tuning knob rather than a villain.

It is hard to appreciate how radical this is without remembering the culture of traditional electronics. Engineers are trained to eliminate noise: filter it, shield it, cool it away. Photosynthetic proteins do not have that option. Their molecules jiggle. Water buffets them. Yet the architecture works, and the jiggling is not merely tolerated; it shapes energy levels and couplings dynamically, turning a static network into a responsive one. When designers talk about mimicking the "noisy protein scaffold," they are not proposing to build fragile quantum computers out of chlorophyll. They are proposing to build *open quantum systems* on purpose: devices where coherence exists, briefly, while the environment is engineered to guide rather than destroy it.

There is also a more mundane but equally important lesson from plants: they are optimized for *variable* light, not just bright noon sunlight. A silicon panel is a marvel of industrial physics, but it is happiest when flooded with photons. In low light, the current falls and the balance of loss mechanisms shifts. Many biological antennas, by contrast, are designed to catch sparse photons and route them efficiently without demanding perfect conditions. For the future of energy, low-light performance matters more than it might seem. Solar power is not only about deserts at midday; it is about cloudy cities, indoor sensors, under-canopy agriculture, wearable devices, and distributed electronics that sip energy rather than guzzle it.

The technological translation begins with the basic mismatch between photosynthesis and photovoltaics. A plant uses excitons to deliver energy to a chemical reaction center that separates charge and stores energy in bonds. A solar cell aims to separate charges in a semiconductor and extract them as current. But excitons are relevant to modern solar technology too, especially in *organic photovoltaics* and in many nanostructured materials where absorption creates tightly bound electron–hole pairs. Even in silicon, excitonic effects can matter near the absorption edge or in engineered nanostructures. If one wants to learn from biology, the question becomes: can we build materials where excitons travel farther, find interfaces more reliably, and hand off their energy without excessive loss?

A useful engineering term here is the *coherence length*: roughly, how far (or over how many sites) a coherent superposition meaningfully extends. In a disordered molecular film at room temperature, coherence length may be small, perhaps only a few molecules. But small is not zero, and small can still help if the transport problem is bottlenecked by traps and energetic dead ends. The dream is not macroscopic coherence spanning millimeters; it is coherent assistance over nanometers, long enough to improve the odds at each step.

One place this thinking has flourished is in bio-inspired nanostructures that imitate a pigment–protein network by arranging chromophores with controlled spacing and coupling. In the lab, researchers can build arrays of dyes, quantum dots, or conjugated polymers and tune their interactions. They can also embed them in matrices that provide controlled vibrational environments. The point is not to recreate chlorophyll's chemistry; it is to recreate the *design logic*: coupled excitonic states, energetic funnels that bias direction, and an environment that does not simply randomize everything but provides a structured noise spectrum.

A concrete example of "structured noise" is the use of vibrational modes—molecular vibrations at specific frequencies—that can resonate with energy gaps between excitonic states. If a vibration matches the gap, it can assist transfer by providing or absorbing the right amount of energy, like a well-timed push on a swing. In biological complexes, such vibronic mixing is one reason the spectroscopy is difficult to interpret, but it is also one reason the transport can be robust. In artificial systems, this suggests a new materials strategy: instead of treating phonons (vibrations in a solid) as mere sources of decoherence, design the phonon spectrum to be helpful.

Another translation of nature's blueprint aims at a different loss mechanism in solar cells: thermalization. In a conventional single-junction cell, a photon with energy above the band gap creates a charge carrier with excess kinetic energy, and that excess is rapidly lost as heat. This is one reason why the Shockley–Queisser limit exists: it is not a technological flaw of silicon but a thermodynamic compromise built into the single-junction architecture. If biology teaches anything, it is that energy should be stepped down through controlled pathways rather than dumped as heat in one gulp. Materials scientists have pursued a quantum-enabled workaround: make one photon do more than one useful thing.

Two related ideas are *multiple exciton generation* (MEG) and *singlet exciton fission*. In MEG, a high-energy photon can produce more than one electron–hole pair in certain nanocrystals, potentially increasing current. In singlet fission, an absorbed photon creates a singlet exciton in an organic molecule, which then splits into two lower-energy triplet excitons on neighboring molecules. In principle, that is a kind of energy multiplication: one photon, two excitations. If those two triplets can be harvested as charges, a solar device could surpass the usual single-junction efficiency ceiling.

Singlet fission is especially interesting for a book about quantum biology because it is not merely a counting trick; it is a story about spin and coherence. Triplet states have different spin configurations than singlets. The intermediate state in singlet fission can be a correlated pair of triplets, and recent work has explored the coherent preparation of these multi-exciton states. The language begins to resemble that used for spin chemistry in living systems: quantum states that are fragile but potentially steerable at room temperature if the molecular architecture is right. Here the inspiration is not that plants do singlet fission in the same way (they may avoid it to prevent harmful triplets), but that nature's molecular precision suggests how to control competing pathways. A plant must route excitation to chemistry while suppressing dangerous side reactions; a solar engineer must route excitation to charges while suppressing recombination.

A historical anecdote helps keep our feet on the ground. The earliest silicon solar cells of the 1950s, developed at Bell Labs, were famously expensive and inefficient by modern standards, but their first compelling use was not rooftops; it was space. The economics made sense when the alternative was no power at all. Photosynthetic organisms inhabit a similar regime: in deep shade or murky water, energy is scarce and every absorbed photon is precious. That scarcity shapes the antenna's priorities. It is no accident that some bacteria have elaborate antenna systems that look, from an engineer's perspective, almost overbuilt. The device is not optimized for peak output under bright sun; it is optimized for survival when photons arrive like occasional coins dropped into a jar.

That perspective shifts how one evaluates "efficiency." A leaf's overall conversion of sunlight into stored chemical energy is not spectacular compared with a good solar panel under ideal conditions. Much of the absorbed energy is used for metabolism, repair, and regulation; plants also dissipate energy deliberately to avoid damage. The remarkable efficiency is local and conditional: within a light-harvesting complex, the probability that an absorbed photon's excitation reaches a reaction center before being wasted can be extremely high. The engineering opportunity lies in capturing that local reliability and combining it with the brute advantages of modern materials.

Consider, for instance, the problem of exciton diffusion length in organic photovoltaic materials. Many organics absorb light strongly, allowing thin films, but their excitons often travel only a few to a few tens of nanometers before recombining. That short diffusion length forces devices to use intricate nanostructured donor–acceptor interfaces so that excitons can reach a charge-separating junction quickly.

But intricate interfaces come with their own penalties: traps, disorder, and recombination sites. If bio-inspired design could extend effective exciton transport—even modestly—it would relax these geometric constraints, enabling devices that are both simpler and more robust.

One approach borrows the idea of an *energy funnel*. In many photosynthetic antennas, pigments are tuned so that excitation energy tends to drift downhill toward the reaction center. This is not a steep waterfall; too steep would slow transport by trapping energy on low sites. It is a gentle slope with local variations, allowing exploration but biasing the result. In artificial systems, engineers can create analogous funnels by grading composition, thickness, or molecular packing so that exciton energies change smoothly across a film. The physics is familiar to semiconductor engineers—bandgap grading in multijunction cells, for example—but the photosynthetic analogy emphasizes the value of a landscape that is neither flat nor cliff-like.

Another approach mimics the protein scaffold's role in maintaining precise pigment geometry. In a leaf, pigments are not embedded randomly; they are held in place with angstrom-level control by a protein that has been refined by evolution. That control sets the coupling strengths between pigments and therefore the splitting of excitonic energy levels. In synthetic materials, self-assembly and supramolecular chemistry aim to achieve a comparable control without a full protein. The payoff is that when chromophores are arranged in a regular way, they can form *exciton delocalization* over multiple units, and delocalization can reduce sensitivity to local disorder. It is a quantum version of redundancy: if the excitation is shared, a single bad site is less catastrophic.

There is a danger, however, in romanticizing coherence as a magic ingredient. One can build highly coherent excitonic systems—at low temperatures, in carefully purified crystals—that still perform poorly as energy devices because they lack a place to send the energy. Photosynthesis works because transport is matched to an irreversible sink: the reaction center acts as a trap that turns excitation into charge separation quickly enough to beat losses. Engineers face an analogous requirement. If an artificial antenna delivers excitation to an interface but the interface fails to separate charge efficiently, the benefit evaporates. Nature's blueprint is not "coherence everywhere"; it is "coherence where it helps, then decisive irreversibility."

That balance between reversible and irreversible steps is one of the most transferable lessons. Coherent motion is reversible: wave-like evolution can in principle run backward. Energy devices need rectification: once the energy reaches the right place, it must be locked into a useful form. In photosynthesis, that lock is chemical: electron transfer steps that quickly separate charges and move them away from one another. In photovoltaics, the lock is electrical: charge separation at a junction followed by carrier extraction before recombination. Either way, the architecture must ensure that the timescales line up: transport faster than loss, trapping fast enough to be decisive, and noise tuned so that it assists exploration without destroying the helpful correlations.

It is also worth emphasizing a more human, less quantum aspect of the story: nature's devices are manufactured in water, at room temperature, from abundant elements, using molecular self-assembly. Even if coherence turned out to be a minor player, the broader blueprint remains valuable. Imagine solar materials that can be grown or printed with minimal energy input, that tolerate defects, and that repair or reconfigure themselves. Photosynthetic organisms do not just harvest light; they maintain their harvesting machinery against damage. Sunlight is not gentle. It creates reactive oxygen species, it breaks bonds, it ages pigments. A leaf survives because it has a maintenance economy: turnover of proteins, protective quenching pathways, dynamic regulation. Translating that to technology might mean materials that are intrinsically photostable, or devices designed for easy regeneration of the active layer.

The most provocative implication of the "quantum edge" is not that future solar panels will look like leaves, but that future energy technologies may be designed less like static machines and more like living networks: adaptable, noisy, and yet controlled. A silicon wafer is a triumph of crystalline order. A photosynthetic antenna is a triumph of functional disorder, where the variations are not a defect but a resource. In between lies a design space we are only beginning to explore.

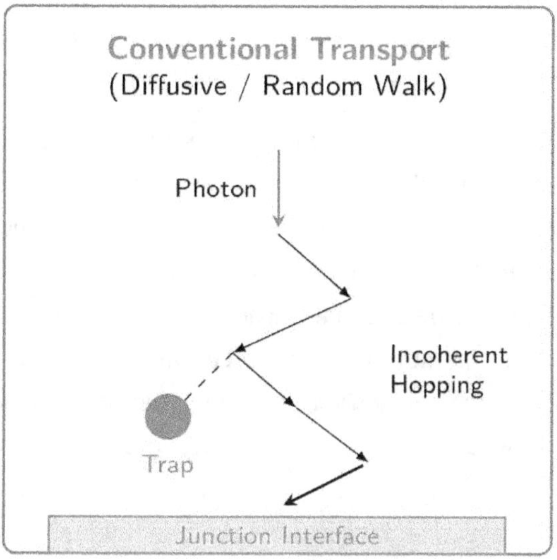

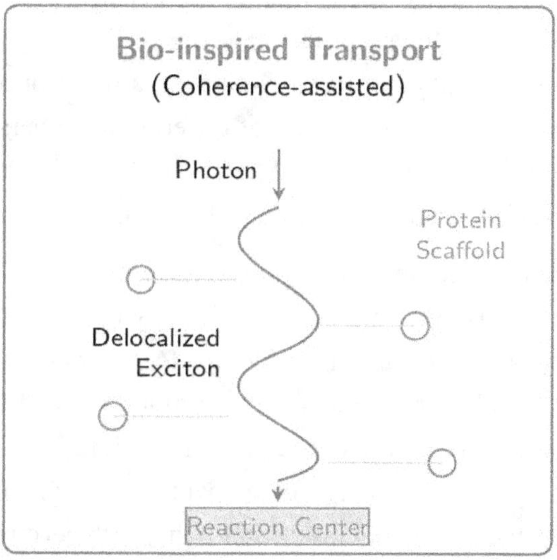

The diagram is a cartoon, but it captures a practical ambition. Conventional devices often treat excitation transport as something that simply happens, then hope it reaches the right interface. Bio-inspired design treats transport as a controllable process: a landscape to sculpt, a network to tune, a set of timescales to match.

Still, one should not confuse *plausible* with *easy*. There are at least three stubborn obstacles. The first is measurement. Coherence and vibronic coupling live on femtosecond timescales and nanometer scales; proving that a particular device benefits from them is far harder than demonstrating that the device works at all. The second is disorder. Industrial fabrication produces variation, and variation can wash out the delicate spectral alignments that make structured noise helpful. The third is integration. Even if a molecular layer performs beautifully in isolation, it must be coupled to electrodes, encapsulated against oxygen and moisture, and manufactured reproducibly.

Yet progress often arrives not as a single leap but as a shift in what engineers believe is worth attempting. A few decades ago, the idea that one might harness "noise" rather than fight it would have sounded perverse. Now it sits comfortably alongside other once-odd ideas: that a solar cell might be a stack of different absorbers, that an interface might be nanostructured like coral, that a device might use molecular self-assembly to do what lithography cannot afford.

Perhaps the most compelling reason to take nature's blueprint seriously is ethical as much as scientific. Energy is not only a technical problem; it is a distribution problem. Technologies that demand ultra-pure crystals, rare elements, or energy-intensive manufacturing tend to concentrate power where capital is already abundant. Photosynthetic systems suggest another pathway: high function from common atoms, built under mild conditions, with performance maintained by architecture rather than brute purity. If a future photocell can be printed from carbon-rich molecules or abundant nanomaterials, tuned to harvest diffuse light efficiently, and engineered so that its inevitable disorder becomes part of its operating principle, it would not merely imitate a leaf. It would inherit a deeper lesson: that sophistication can arise from managing imperfection rather than eliminating it.

9.2 The Quantum Pharmacopoeia

A pill is a vote of confidence in invisible events. Swallow it, and within minutes millions of molecules are making decisions in the dark: docking, twisting, protonating, deprotonating, reacting. The everyday language of pharmacology—"blocks," "activates," "binds"—is reassuringly mechanical, as if a drug were a key in a lock. The uncomfortable truth is that many of the most important steps are not mechanical at all. They are probabilistic, shaped by thermal agitation, water, and the strange willingness of light particles—especially the lightest of all, hydrogen—to ignore barriers.

The word *tunneling* sounds like science fiction, but the underlying idea is almost homely. In classical physics, if a ball does not have enough energy to roll over a hill, it stays put. Quantum mechanics allows a different outcome: if the ball is described by a wave, a little of that wave can leak through the hill, giving a small but real chance of appearing on the other side. For electrons, tunneling is a standard tool of technology; it underlies the scanning tunneling microscope and the operation of some electronic devices. In biochemistry, the surprising part is not that tunneling is possible, but that it can matter at room temperature, in water, in a protein that looks more like a trembling sponge than a precision instrument.

Hydrogen is the natural suspect. Because it is so light, its quantum wavefunction is relatively spread out, and its tunneling probability can be large compared with heavier atoms. But biochemistry is full of hydrogen motion: protons hopping between acids and bases, hydrogens being transferred in redox reactions, hydrogen bonds forming and breaking in DNA base pairs. If tunneling gives even a modest boost to a crucial step, evolution has had billions of years to exploit it.

The practical hook for medicine is that tunneling is not merely an exotic curiosity; it is a lever. Change the mass of the transferred particle and you change the tunneling probability. That is where isotopes enter the story.

Hydrogen has a heavier sibling, deuterium, in which the nucleus contains a proton and a neutron rather than a proton alone. Chemically, deuterium behaves almost like hydrogen: it forms bonds in the same places, with nearly the same geometry. But physically it is twice as heavy, and that matters. A heavier particle has a shorter de Broglie wavelength, a more compact quantum wavefunction, and a lower probability of tunneling through the same barrier. Even when the motion is not pure tunneling, the heavier mass shifts vibrational frequencies and zero-point energies. The net effect shows up in a laboratory measurement that biochemists have relied on for decades: the *kinetic isotope effect*.

The kinetic isotope effect is a mouthful that hides a simple idea. Suppose an enzyme transfers a hydrogen atom during a reaction. Replace that hydrogen with deuterium and repeat the experiment. If the reaction slows noticeably, the bond to hydrogen is involved in the rate-limiting step—the bottleneck. The magnitude of the slowdown can be a clue to mechanism. Large effects, especially with distinctive temperature dependence, often point toward tunneling or at least toward a reaction coordinate where quantum motion cannot be ignored. Over the years, the isotope effect has become a kind of fingerprinting tool: it helps chemists decide whether an enzyme is moving hydrogens by straightforward over-the-barrier motion, by tunneling, or by a mixture of both.

A historical anecdote makes the leap from mechanistic clue to therapeutic strategy feel less abrupt. In the 1930s, Harold Urey and colleagues isolated deuterium and made "heavy water," and within a few years biologists were feeding it to organisms, curious about the consequences. Heavy water is not a drug in any helpful sense—in large amounts it is harmful—but those early experiments established the most important practical fact about isotopes: swapping hydrogen for deuterium can change biology without changing chemistry in any obvious way. Enzymes still bind. Receptors still recognize. Yet rates shift, subtly or dramatically, because the quantum mechanics of a bond has been altered.

That observation is the seed of a modern pharmaceutical idea: *deuterated drugs*. The basic strategy is disarmingly modest. Take an existing drug molecule and replace one or more hydrogens with deuterium at positions that are known to be metabolically vulnerable. Many drugs are broken down by enzymes in the liver, often by cytochrome P450s, through steps that involve hydrogen abstraction. If deuterium substitution slows that step, the drug may be cleared more slowly, leading to a longer half-life, steadier blood levels, fewer frequent doses, or fewer toxic metabolites. No new receptor has to be discovered; sometimes the medicinal chemist simply rebalances the pharmacokinetics—how the body handles the drug—by nudging a quantum bottleneck.

The most widely discussed clinical example is deutetrabenazine, approved in the United States in 2017 for chorea associated with Huntington's disease and later for tardive dyskinesia. It is a deuterated version of tetrabenazine, a drug that affects vesicular monoamine transport. Deuteration did not create a new mode of action; it primarily altered metabolism, helping produce more stable exposure in the body and allowing different dosing. This is a rare case where a phrase as abstract as "kinetic isotope effect" is quietly present in a prescription decision.

Deuterated drugs, however, are not a magic wand, and the limitations are instructive because they illuminate the boundary between quantum and clinical reality. Slowing a metabolic step can be helpful, but it can also shift the balance toward other pathways, producing unexpected metabolites. Deuteration can change binding affinity slightly in some contexts, not because deuterium has different chemistry, but because altered vibrational energies can subtly reshape conformational preferences and hydrogen-bond strengths. Patent law and pricing controversies hover in the background too: a deuterated version of an existing drug can be medically useful, but it can also be a business strategy. The physics does not decide the ethics, but it does provide a rare case where quantum effects become part of a public conversation about what counts as a "new" medicine.

The deeper scientific point is that deuteration works best when there is a genuine bottleneck involving hydrogen motion. To understand that bottleneck, enzymologists often talk about a reaction coordinate: a collective description of how the system moves from reactants to products. In many reactions, the crucial step is a coupled dance of electrons and protons. That coupling has a formal name, *proton-coupled electron transfer* (PCET), and it sits at the crossroads of chemistry, biology, and energy technology.

PCET can be imagined as the smallest possible relay race. An electron moves, a proton moves, and the two motions are linked because moving one without the other can be energetically costly. Sometimes the electron moves first, creating a charged intermediate that then pulls a proton along. Sometimes the proton moves first. In many biologically relevant cases, the most efficient route is *concerted*: electron and proton shift together in a single coordinated step, avoiding high-energy charged intermediates. When the proton is light and the transfer distance is short, tunneling can contribute significantly to the rate. The protein environment is not a passive container here; it shapes distances, electric fields, and the alignment of donor–acceptor groups. A protein can, in effect, tune the barrier width for proton motion.

That tuning does not look like the rigid precision of a machine. Proteins breathe. Side chains flicker between conformations. Water molecules come and go. This motion can be an ally. A useful way to picture it is to imagine that the barrier for a proton is not fixed; it fluctuates. Every so often, by thermal motion, the donor and acceptor come unusually close, the hydrogen bond briefly straightens, and the barrier becomes thin enough for tunneling to spike. The overall reaction rate then depends on two things: the probability of those favorable conformations, and the quantum probability of transfer when they occur. In this view, the protein is not tunneling *instead of* classical dynamics; it is using classical dynamics to create quantum opportunities.

Drug designers have begun to treat this as more than an intellectual curiosity. Many drugs target enzymes. If an enzyme's catalytic power depends on arranging a precise donor–acceptor distance for hydrogen transfer, then a small molecule that disrupts that arrangement—even without blocking the active site outright—could be an inhibitor. Conversely, if an enzyme is deficient because it cannot reach those favorable conformations, one might imagine designing a small molecule that stabilizes them, acting as an activator. This is a more delicate form of control than simply plugging the active site. It moves pharmacology toward *dynamic* allostery: changing the probabilities of motions that the enzyme needs to harness quantum transfer.

This is also where the romance of quantum biology needs discipline. Not every large isotope effect is tunneling, and not every enzymatic hydrogen transfer is meaningfully quantum in a way that can be exploited therapeutically. Sometimes deuteration fails to improve a drug because the rate-limiting step is not hydrogen abstraction at all, or because the body compensates through other clearance routes. Sometimes an isotope effect that is dramatic in a purified enzyme disappears in cells because transport and binding dominate. The useful attitude is neither dismissal nor enchantment but a kind of mechanistic realism: identify where hydrogen motion is the bottleneck, measure it, and then decide whether isotopes or dynamics are an actionable handle.

The most provocative application of tunneling to medicine is not about drug clearance, but about mutation. DNA is stabilized by hydrogen bonds between bases. Those hydrogen bonds are not just static attractions; they involve protons sitting in asymmetric double wells: positions that strongly favor the standard Watson–Crick pairing, with less favorable alternative positions corresponding to rare *tautomers*—forms of the bases that can mispair during replication. The idea that proton motion might create mutagenic tautomers is old. It appears in early speculations by Per-Olov Löwdin and even in hints in the Watson–Crick era: if a proton could jump, a base might briefly masquerade as another, and a replication enzyme might incorporate the wrong partner, seeding a point mutation.

The modern twist is that quantum mechanics makes such jumps more plausible than classical thermal hopping would suggest. A proton could tunnel between positions, at least in principle. If that tunneling happened often enough and the tautomer persisted long enough, it could contribute to the baseline mutation rate.

Here the evidence becomes contentious, and the controversy is instructive precisely because it shows how hard it is to translate a quantum possibility into a biological fact. The probability that a tautomer exists is not an intrinsic property of an isolated base pair; it depends strongly on the environment: the local electric fields, the presence of water, the constraints of the double helix, and the active site of polymerases that copy DNA. Different theoretical treatments that include these environmental effects in different ways can yield very different conclusions. Some open-quantum-systems models, particularly for G–C pairs in certain contexts, suggest higher equilibrium tautomer populations than one might naively expect. Other modeling, often focusing on A–T pairs and using different treatments of the surrounding environment, finds tunneling-generated tautomers to be vanishingly rare. Both sets of results can be internally consistent because the system is exquisitely sensitive to assumptions about solvation and constraint. A proton in a hydrogen bond is not a free particle; it is a quantum object embedded in a noisy, structured, charged environment.

Even if tunneling contributes to rare tautomers, the leap to drug design is not straightforward. One can imagine, in a far future, molecules that stabilize DNA in ways that suppress mutagenic tautomerization, perhaps reducing mutation rates in certain contexts. One can also imagine the opposite, for cancer therapy: increasing replication stress or mutation rates selectively in tumor cells. But the immediate relevance is humbler and more valuable: mechanistic understanding of mutagenesis informs risk assessment for chemicals, interpretation of mutation spectra, and the development of polymerase inhibitors and antiviral nucleoside analogs. A good quantum model of proton motion in base pairs would not hand medicine a new pill overnight; it would sharpen the map of where errors arise.

A cautionary tale sits nearby, tempting precisely because it offers a seductive diagnostic fantasy: the idea that biology might *sense* molecular vibrations quantum mechanically, and that medicine could exploit that sensing. The most famous version is vibrational olfaction, a hypothesis that odor receptors detect not only shape but also vibrational spectra, perhaps through inelastic electron tunneling.

There have been experiments suggesting that fruit flies can distinguish isotopomers—molecules that have the same shape but different isotope composition, and therefore different vibrational frequencies. It is an intriguing result because it hints at a sensory pathway that is, in a strict sense, quantum-aware.

Yet receptor-level critiques have been persistent: the structural biology and biophysics of odor receptors make it difficult to see how such a mechanism would operate robustly in the messy chemical environment of the nose, and many odor phenomena are well explained by more classical models involving binding and activation. The dispute is not merely academic. If vibrational sensing were a general biological capacity, one could imagine a "quantum diagnostic" that reads disease markers by their vibrational fingerprints. If it is not, then such a diagnostic is more likely to be a mirage. The responsible stance is to let the controversy do its job: it warns that a quantum mechanism can be clever on paper yet implausible in the full biological context.

The pharmacological uses of isotopes, by contrast, do not require biology to be a quantum sensor. They require only that bond-breaking and bond-making rates depend on mass in predictable ways, which they do. That modesty is why deuteration has already entered the clinic while vibrational diagnostics remain speculative.

Still, the most exciting future possibilities may not be about swapping hydrogen for deuterium, but about targeting the *protein motions* that create tunneling opportunities. Enzymes that rely on hydrogen transfer often exhibit networks of residues whose coordinated motion modulates the donor–acceptor distance and the active-site electrostatics. A drug that binds far from the active site can alter these networks, changing the distribution of conformations and therefore the effective rate of transfer. This is, in a sense, a way of practicing quantum pharmacology without needing quantum control: the drug acts classically, reshaping a fluctuating landscape in which quantum transfer is occasionally decisive.

There is also a philosophical payoff that matters for medicine. People often imagine quantum effects as either negligible or overwhelming. The reality is more granular. A proton's tunneling event may be rare, but if it is the bottleneck in an enzymatic pathway, it controls the flow of metabolites. A deuterium substitution may change a rate by a factor of two or five, which sounds small, until one remembers that such factors can separate a safe dose from a toxic one, or make the difference between once-a-day and three-times-a-day dosing. In pharmacology, small shifts in rate constants can echo through feedback loops and clearance mechanisms into large differences in outcome.

The quantum pharmacopoeia, at its best, is not a cabinet of mystical remedies. It is a style of attention to the lightest atoms, the shortest timescales, and the quiet places where probability is destiny. A single proton, hesitating at a barrier inside an enzyme, can decide whether a drug is cleared quickly or lingers long enough to help, whether a metabolic byproduct accumulates or disappears, whether a molecular handshake is fleeting or sustained. Medicine has always depended on the invisible. Quantum mechanics simply adds a sharper vocabulary for the smallest decisions.

9.3 A Synthetic Sixth Sense

A compass feels like a solved problem until you try to build one without metal, without a battery, and without any moving parts. Migratory birds do exactly that, and they do it while flying through weather, over oceans, and under night skies where landmarks are sparse. The deeper surprise is not merely that birds can sense the Earth's magnetic field—many animals respond to magnetic cues in one way or another—but that at least one leading explanation makes the sensing a matter of *spin*.

Spin is a quantum property that resists everyday analogy. It behaves a little like an intrinsic angular momentum and a little like a label that determines how a particle responds to magnetic fields. An electron has spin, and that spin can point in superpositions rather than simply "up" or "down." In most materials at room temperature, such delicate orientation information is rapidly scrambled by collisions and thermal motion. Yet in the avian compass hypothesis, spin information lasts long enough to influence chemistry, and chemistry lasts long enough to influence perception.

The proposal, now known as the *radical-pair mechanism*, has the kind of elegance that makes scientists both excited and cautious. It links three ordinary ingredients—light, a protein, and weak magnetic fields—using quantum dynamics that are subtle but testable. The protein most often discussed is cryptochrome, a blue-light photoreceptor found in plants and animals. Cryptochromes contain a light-absorbing cofactor (a flavin) and, in many versions, a chain of amino acids (often tryptophans) that can pass an electron along in a series of steps. Under illumination, an electron can be transferred from a donor residue to the excited flavin, creating two separated unpaired electrons: one on the flavin and one on the donor side. Each unpaired electron is a *radical*. Together they form a *radical pair*.

Why is a radical pair special? Because two unpaired electron spins can be correlated. At birth, the pair is typically created in a specific combined spin state, often a *singlet* state (spins anti-aligned in a quantum-correlated way) or a *triplet* state (with different aligned configurations). The key is that singlet and triplet are not just labels; they are chemically consequential. Many radical pairs recombine back to a closed-shell product only efficiently from the singlet state. If the pair is in a triplet state, recombination may be slowed or diverted, sending the chemistry down a different path.

Enter the Earth's magnetic field, which at the surface is on the order of tens of microtesla. To a physicist used to MRI magnets, that is almost nothing. Yet electron spins are exquisitely sensitive. Even such a weak field can slightly shift the relative energies of spin states and change the way singlet–triplet components interconvert over time. The radical-pair mechanism proposes that the geomagnetic field nudges the singlet–triplet mixing, which nudges the fraction of reactions that end in one product versus another. The output of the sensor is not a needle pointing north; it is a *chemical yield*.

This is, in a literal sense, a sixth sense: a new sensory channel built out of chemistry. It is also a striking answer to a practical engineering question: how do you measure a weak magnetic field at room temperature without cryogenics or large coils? Make a molecule do it.

A historical anecdote helps show how unexpected this idea was. In the late 20th century, Klaus Schulten and colleagues developed early theoretical frameworks suggesting that radical pairs could be magnetic-field sensitive, and that biology might exploit this in a light-dependent compass. At the time, the concept competed with other hypotheses, including magnetite-based sensing, in which tiny crystals act like biological compass needles. The magnetite story remains relevant in some species and contexts, but the radical-pair idea had a distinct and testable signature: *light dependence*. If the compass depends on a photochemical reaction, then changing illumination spectrum or intensity should affect orientation. Behavioral experiments indeed showed that certain birds' orientation can depend on light wavelength, a clue that made cryptochrome feel less like an arbitrary candidate and more like a plausible culprit.

Mechanisms are judged by their vulnerabilities. The radical-pair mechanism has one that is immediately obvious: decoherence. A radical pair lives in a protein, in water, at body temperature. It is buffeted by molecular motion and by magnetic fields from nearby atomic nuclei. Those nuclear fields are not an annoyance; they are central to the mechanism.

The unpaired electron spins interact with nearby nuclear spins through *hyperfine coupling*. Each nucleus—hydrogen, nitrogen, carbon—acts like a tiny magnet. The pattern of couplings is like a local magnetic fingerprint that differs depending on exactly where the unpaired electron resides. Hyperfine interactions are what allow singlet and triplet character to mix even in zero applied field; the Earth's field then adds a bias, changing the mixing dynamics just enough to alter yields. In other words, the radical pair is not a delicate quantum system that requires isolation; it is a system in which controlled, structured interactions with the environment generate the signal.

That still leaves a formidable requirement: the radical pair must survive long enough for the weak geomagnetic field to matter. If recombination is too fast, there is no time for singlet–triplet interconversion. If spin relaxation is too fast, the initial correlations are erased. The fact that birds seem to manage suggests that cryptochrome (or whatever molecular machinery is involved) has been tuned by evolution to sit in a narrow window: long-lived enough to be sensitive, short-lived enough to be reset and responsive.

The state of the evidence is not a neat courtroom verdict, but several pieces have become hard to ignore. In 2016, P. J. Hore and Henrik Mouritsen wrote a widely read synthesis of the radical-pair mechanism that emphasized what sets sensitivity: hyperfine couplings, relaxation times, recombination pathways, and the need for directional information—an inclination compass rather than a polarity compass, matching bird behavior. More recently, in vitro experiments have strengthened the bridge between idea and molecule. Work including Jingjing Xu and Lauren Jarocha (2021) reported magnetic-field effects in cryptochrome 4 from the migratory European robin, with comparative sensitivity versus non-migratory species and mechanistic hints from mutations. Those experiments do not prove that a bird sees magnetism *only* via that protein in its eye, but they show that the protein can, in principle, do the physics.

For engineers, "in principle" is an invitation. If a protein can transduce microtesla fields into a chemical signal at room temperature, then a device might do the same. The prospect is not merely a novelty compass; it is navigation without satellites, sensing in environments where GPS is jammed or unavailable, and compact magnetic detectors that work without cryogenic cooling. It is also a chance to build sensors that operate in regimes where conventional magnetometers struggle: tiny volumes, low power, and integration with optics and microfluidics.

To imagine a radical-pair sensor, it helps to translate the bird's invisible biochemistry into a block diagram. There must be an input that creates radical pairs (light), a reaction chamber that hosts them (a cryptochrome-like molecule or synthetic analog), a way the magnetic field changes the spin dynamics (through Zeeman shifts and hyperfine couplings), and a readout that converts chemical yield into a measurable signal (fluorescence, absorption, current, or a downstream catalytic change). The sensor is, in effect, a chemical amplifier whose gain depends on spin coherence.

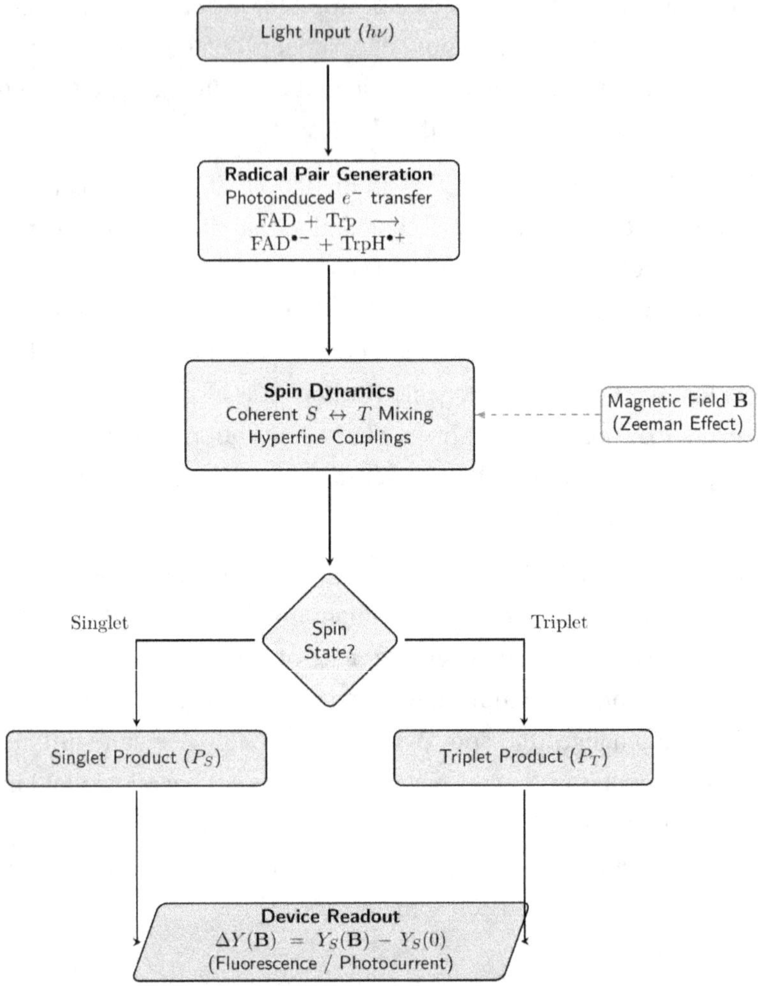

A diagram like this makes the radical-pair sensor look almost straightforward: shine light, get a magnetic-field-dependent signal. The devil is in the word "dependent." The yield changes caused by microtesla fields are typically small. Nature can afford small signals because it can average over time and because biology is built to amplify subtle chemical differences through neural processing. A human-made device must be explicit about its amplification: longer averaging, higher radical-pair densities, clever optical cavities, lock-in detection, or chemical cascades.

One engineering decision comes immediately: what plays the role of cryptochrome? One can try to use cryptochrome itself, perhaps immobilized on a surface, embedded in a polymer, or expressed in engineered cells. That path has an appealing directness, and it leverages evolution's tuning. But proteins are fussy. They denature, they require specific environments, and they can be difficult to integrate into scalable manufacturing. The alternative is to design synthetic donor–acceptor molecules that create radical pairs with tailored hyperfine couplings and long lifetimes, effectively building a "cryptochrome-like" reaction chamber from scratch.

Either approach forces a deep confrontation with the physics of open quantum systems, but in a different dialect than quantum computing. A quantum computer tries to preserve coherence as long as possible and to isolate qubits from uncontrolled noise. A radical-pair sensor is happier with *some* noise, because hyperfine couplings and environmental interactions generate the very singlet–triplet mixing that makes magnetic sensitivity possible. The goal is not pristine isolation; it is a controlled balance between coherence (to remain field-sensitive) and relaxation (to reset and produce a usable yield). The phrase "room temperature" is not a handicap in this context; it is part of the design brief.

Directionality is another subtle requirement. Birds do not merely detect field strength; they behave as if they sense the *inclination* of the magnetic field lines, distinguishing poleward from equatorward directions without necessarily distinguishing north from south. In radical-pair terms, directionality arises because the hyperfine coupling tensor is anisotropic: the molecule has an orientation, and the magnetic field's effect depends on how it is aligned relative to that molecular frame. A sensor that uses randomly oriented molecules in solution may detect field strength but lose directional information. A sensor that aligns molecules—by crystallization, surface attachment, liquid crystals, or stretched polymers—can, in principle, recover directionality. This is where biology offers another quiet lesson: proteins are not floating randomly in a beaker; they are embedded in structured tissues with preferred orientations.

A concrete application brings the stakes into focus. Modern navigation, from ships to smartphones, leans heavily on satellite signals and on inertial sensors that drift over time. In contested environments, GPS can be jammed or spoofed. Submarines and deep underground operations cannot rely on it at all. High-end alternatives exist—atomic magnetometers, SQUIDs—but they can be power-hungry, bulky, or require special operating conditions. A radical-pair sensor that works at room temperature, in a small package, with optical readout, would occupy a unique niche: not necessarily the most sensitive magnetometer in the world, but one that is compact, potentially cheap, and compatible with distributed sensing.

There is also a biomedical and environmental angle. Magnetic fields are used in medical imaging and stimulation, but measuring weak magnetic signatures in biological settings is hard. One can imagine, at least in the long term, implantable or wearable magnetosensors that monitor magnetic nanoparticles used in targeted drug delivery, or that map subtle field changes near devices. The radical-pair approach is not obviously the best for these tasks, but it expands the palette: it suggests sensors that can be soft, molecular, and integrated with chemistry.

The radical-pair mechanism also resonates with an unexpected neighbor: the quantum engineering of energy materials. In singlet exciton fission, discussed earlier in the context of solar technologies, a key intermediate can be a correlated triplet pair whose spin state and coherence influence what happens next. Researchers have begun to treat the spin character of that intermediate not as an inconvenient detail but as a handle for controlling yields. The conceptual tools overlap: singlet–triplet manifolds, spin relaxation, entanglement versus classical mixtures, and the problem of readout. A radical-pair magnetosensor and a singlet-fission solar layer are not the same device, but they live in the same conceptual neighborhood: chemistry where spin is not merely bookkeeping, but function.

Skepticism remains healthy here. Not all observed magnetic effects in proteins necessarily imply a navigational compass, and not all in vitro sensitivity implies in vivo relevance. The retina is a complex place, and animals are clever at integrating cues. There may be multiple magnetic sensing pathways, including magnetite-based mechanisms in some species, and the precise neural encoding of a radical-pair signal is still an active area of debate. But the engineering impulse does not require every biological detail to be settled. It requires only that a reproducible, field-dependent chemical yield can be generated and measured, and that the signal can be scaled into a device.

There is something emotionally disarming about this prospect. A compass usually feels like an instrument that imposes its certainty on the world: north is there, trust the needle. The radical-pair compass is more like a whisper, a statistical bias in competing chemical futures. It says that measurement does not always require a rigid mechanism. Sometimes it is enough to set up a system whose internal probabilities are slightly tilted by the world, and then to read out that tilt patiently.

If technology manages to build a practical sensor from this principle, it will not look like a bird's eye, and it will not confer animal intuition on machines. It will do something subtler and, in its own way, more profound: it will turn the Earth's magnetic field into a variable in a molecular reaction, and then into a signal that circuits can understand. That is a kind of translation between domains—field to spin to chemistry to information—carried out at room temperature by the quiet arithmetic of quantum probability.

www.ingramcontent.com/pod-product-compliance
Lightning Source LLC
LaVergne TN
LVHW022202200926
848481LV00006B/1817
9798896652861